Functional integration
and its applications

Functional integration and its applications

Proceedings of the International Conference
held at Cumberland Lodge, Windsor Great Park,
London, in April 1974

EDITED BY
A. M. ARTHURS

CLARENDON PRESS · OXFORD
1975

Oxford University Press, Ely House, London W. 1
GLASGOW NEW YORK TORONTO MELBOURNE WELLINGTON
CAPE TOWN IBADAN NAIROBI DAR ES SALAAM LUSAKA ADDIS ABABA
DELHI BOMBAY CALCUTTA MADRAS KARACHI LAHORE DACCA
KUALA LUMPUR SINGAPORE HONG KONG TOKYO

ISBN 0 19 853346 2

©OXFORD UNIVERSITY PRESS 1975

All rights reserved. No part of this publication may be reproduced, stored in a retrieval system, or transmitted, in any form or by any means, electronic, mechanical, photocopying, recording or otherwise, without the prior permission of Oxford University Press

PRINTED IN GREAT BRITAIN
BY J. W. ARROWSMITH LTD.,
BRISTOL, ENGLAND

Foreword

THE articles in this book represent the formal part of the programme at the International Conference on Functional Integration and its Applications held in April 1974, at Cumberland Lodge, Windsor Great Park, London. The aim of this Conference was to bring together mathematicians and physicists concerned with this many-sided area of research, and the discussions ranged over mathematical and computational aspects of functional integration, as well as its applications in several branches of physics including field theory, quantum mechanics, and statistical mechanics.

We felt from the beginning that the publication of the more formal part of the Conference Proceedings would provide a valuable record and guide both for experts and advanced students. In addition, we hope that the present volume will stimulate further investigations in this particular field of mathematics and its applications. The excellent co-operation of the speakers at the Conference has made it possible to publish at a relatively early date.

Financial support for the Conference was provided by the United Kingdom Science Research Council. Its scientific planning was the work of a committee consisting of S. F. Edwards, M. R. Bhagavan, and the present editor.

University of York A. M. A.
September 1974

Contents

LIST OF PARTICIPANTS — ix

1. SOME INEQUALITIES FOR GAUSSIAN MEASURES AND THE LONG-RANGE ORDER OF THE ONE-DIMENSIONAL PLASMA — 1
   H. J. BRASCAMP AND E. H. LIEB
   1.1. Introduction. 1.2. Basic concavity theorem. 1.3. Application of Theorem 1.1 to Gaussian measure. 1.4. The one-dimensional plasma.

2. ASYMPTOTIC EVALUATION OF CERTAIN WIENER INTEGRALS FOR LARGE TIME — 15
   M. D. DONSKER AND S. R. S. VARADHAN
   2.1. Introduction. 2.2. The upper estimate. 2.3. The lower estimate. 2.4. The equality of the estimates.

3. COVARIANT SCHRÖDINGER EQUATIONS — 34
   J. S. DOWKER
   3.1. Deriving the equations. 3.2. Solving the equations. 3.3. Conclusion. 3.4. Appendix.

4. FUNCTIONAL PROBLEMS IN THE THEORY OF POLYMERS — 53
   S. F. EDWARDS
   4.1. Introduction. 4.2. A problem from dilute solution theory. 4.3. Rubber elasticity. 4.4. Entanglements of chains.

5. MEASURES ON INFINITE-DIMENSIONAL MANIFOLDS — 60
   K. D. ELWORTHY
   5.1. Introduction. 5.2. Abstract Wiener manifolds. 5.3. Construction by stochastic differential equations. 5.4. General remarks.

6. THE FREE EUCLIDEAN PROCA AND ELECTROMAGNETIC FIELDS — 69
   L. GROSS
   6.1. Introduction. 6.2. The Proca field. 6.3. The electromagnetic field.

7. THE DESIGN OF FUTURE COMPUTING MACHINERY FOR FUNCTIONAL INTEGRATION — 83
   J. M. HAMMERSLEY

8. SOME PROBABILISTIC ASPECTS OF SCATTERING THEORY — 87
   M. KAC AND P. VAN MOERBEKE
   8.1. Introduction. 8.2. A trace formula. 8.3. Introducing the Konteweg–de Vries flow. 8.4. Konteweg–de Vries flow continued. 8.5. A question of isospectrality. 8.6. A discrete analogue.

9. HOW TO MAKE A HEAT BATH ... 97
J. T. LEWIS AND L. C. THOMAS

9.1. Introduction. 9.2. Linear stochastic differential equations. 9.3. Vector-valued processes. 9.4. Dilating semigroups of contractions. 9.5. The Ford–Kac–Mazur model. 9.6. A string model. 9.7. Lax–Phillips structures. 9.8. Remarks.

10. FUNCTIONAL INTEGRALS AND LOCAL MANY-BODY PROBLEMS: LOCALIZED MOMENTS AND SMALL PARTICLES ... 124
B. MÜHLSCHLEGEL

10.1. Method of functional integrals for general many-body systems. 10.2. The localized magnetic moment problem. 10.3. Zero-dimensional superconductors.

11. REMARKS ON MARKOV FIELD EQUATIONS ... 136
E. NELSON

11.1. Introduction. 11.2. The lattice approximation. 11.3. The Markov field equation. 11.4. Speculations.

12. CAUSTICS AND MULTIVALUEDNESS: TWO RESULTS OF ADDING PATH AMPLITUDES ... 144
L. SCHULMAN

12.1. Propagators for multivalued wave functions. 12.2. Caustics in electron optics.

13. FUNCTIONAL INTEGRATION AND INTERACTING QUANTUM FIELDS ... 157
I. E. SEGAL

13.1. Introduction. 13.2. Renormalized powers of quantized distributions. 13.3. Invariant integration in solution manifolds of relativistic equations.

14. DEFINITIONS AND SELECTED APPLICATIONS OF FEYNMAN-TYPE INTEGRALS ... 169
J. TARSKI

14.1. Examples and definitions. 14.2. Some problems of integration. 14.3. Applications.

15. PROBLEMS IN QUANTUM GRAVITY ... 181
J. G. TAYLOR

15.1. Introduction. 15.2. Divergences and ambiguities. 15.3. Choice of variables. 15.4. Operator ordering. 15.5. The loop expansion method. 15.6. Counter terms in the loop expansion. 15.7. Conclusion.

AUTHOR INDEX ... 189

SUBJECT INDEX ... 192

List of Participants

ARTHURS, A. M.	Department of Mathematics, University of York, York YO1 5DD, England.
BHAGAVAN, M. R.	Department of Mathematics, Bedford College, London NW1 4NS, England.
BLOMBERG, C.	Department of Theoretical Physics, Royal Institute of Technology, 10044 Stockholm 70, Sweden.
DAVIES, E. B.	St. John's College, Oxford OX1 3JP, England.
DEWITT, C.	Department of Astronomy, University of Texas, Austin, Texas 78712, U.S.A.
DINGLE, R. B.	Department of Theoretical Physics, University of St Andrews, St Andrews, Scotland.
DONSKER, M. D.	Courant Institute of Mathematical Sciences, New York University, 251 Mercer Street, New York, NY 10012, U.S.A.
DOWKER, J. S.	Department of Theoretical Physics, The University, Manchester M13 9PL, England.
EDWARDS, S. F.	Cavendish Laboratory, Cambridge. Now at Science Research Council, State House, High Holborn, London WC1R 4TA, England.
EELLS, J.	Mathematics Institute, University of Warwick, Coventry CV4 7AL, England.
ELWORTHY, K. D.	Mathematics Institute, University of Warwick, Coventry CV4 7AL, England.
GINIBRE, J.	Physique Théorique, Université de Paris, 91405 Orsay, France.
GROSS, L.	Department of Mathematics, Cornell University, Ithaca, NY 14850, U.S.A.
HAMMERSLEY, J. M.	Trinity College, Oxford, England.
KAC, M.	The Rockefeller University, New York, NY 10021, U.S.A.
KATZ, A.	L.T.V. Aerospace Corporation, Dallas, Texas, U.S.A.
KEITER, H.	Institut für Theoretische Physik, Universität zu Köln, Universitätstr 14, 5 Köln 41, Germany.
LEWIS, J. T.	Dublin Institute for Advanced Studies, 10 Burlington Road, Dublin 4, Ireland.
LIEB, E.	Department of Mathematics, MIT, Cambridge, Massachusetts 02139, U.S.A.
LUKES, T.	Department of Applied Mathematics and Mathematical Physics, University College, Cardiff CF1 1XL, Wales.
MARTIN, J. L.	Department of Physics, King's College, London WC2R 2LS, England.
MÜHLSCHLEGEL, B.	Institut für Theoretische Physik, Universität zu Köln, Universitätstr 14, Köln 41, Germany.
NELSON, E.	Department of Mathematics, Princeton University, Princeton, NJ 08540, U.S.A.
NORCLIFFE, A.	Department of Mathematics and Computing Sciences, Sheffield Polytechnic, Sheffield S1 1WB, England.
PAPADOPOULOS, G.	School of Mathematics, The University, Leeds LS2 9JT, England.
PLYMEN, R.	Department of Mathematics, Bedford College, London NW1 4NS, England.
RINGROSE, J. R.	School of Mathematics, The University, Newcastle-upon-Tyne NE1 7RU, England.

List of Participants

ROBERTS, M. J. — Department of Physics, The University, Stirling FK9 4LA, Scotland.
SAMATHIYAKANIT, V. — Department of Physics, Chulalongkorn University, Bangkok, Thailand.
SCHMIDT, K. — Mathematics Institute, University of Warwick, Coventry CV4 7AL, England.
SCHULMAN, L. S. — Department of Physics, Indiana University, Bloomington, Indiana 47001, U.S.A.
SEGAL, I. — Department of Mathematics, MIT, Cambridge, Massachusetts 02139, U.S.A.
SHERRINGTON, D. — Department of Physics, Imperial College, London SW7 2BZ, England.
STREATER, R. F. — Department of Mathematics, Bedford College, London NW1 4NS, England.
SYMANZIK, K. — Deutsches Elektronen-Synchrotron, DESY, Notkestieg 1, 2 Hamburg 52, Germany.
TARSKI, J. — Institut für Theoretische Physik, Universität Hamburg, Luruper Chaussee 149, 2 Hamburg 50, Germany.
TAYLOR, J. G. — Department of Mathematics, King's College, London WC2 2LS, England.
TAYLOR, S. J. — Department of Mathematics, Westfield College, London NW3 7ST, England.
THOULESS, D. J. — Department of Mathematical Physics, The University, Birmingham B15 2TT, England.
ZITTARTZ, J. — Institut für Theoretische Physik, Universität zu Köln, Universitätstr 14, 5 Köln 41, Germany.

1. Some inequalities for Gaussian measures and the long-range order of the one-dimensional plasma

H. J. BRASCAMP AND E. H. LIEB

1.1. Introduction

THE following is a preliminary report on some recent work, the full details of which will be published elsewhere. We have come across some inequalities about integrals and moments of log concave functions which hold in the multidimensional case and which are useful in obtaining estimates for multidimensional modified Gaussian measures. By making a small jump (we shall not go into the technical details) from the finite to the infinite dimensional case, upper and lower bounds to certain types of functional integrals can be obtained. As a non-trivial application of the latter we shall, for the first time, prove that the one-dimensional one-component quantum-mechanical plasma has long-range order when the interaction is strong enough. In other words, the Wigner lattice can exist, in one dimension at least. As another application we shall prove a log concavity theorem about the fundamental solution (Green's function) of the diffusion equation.

1.2. Basic concavity theorem

We begin with a theorem (Theorem 1.1) which, to the best of our knowledge, is new and which constitutes the basis of all our other inequalities.

DEFINITION 1.1. A function F from $\mathbf{R}^n$ to $\mathbf{R}$ is a *log concave function* if $F(x) \geq 0$, $\forall x \in \mathbf{R}^n$, and $F(x)^\lambda F(y)^{1-\lambda} \leq F[\lambda x + (1-\lambda)y]$, $\forall x, y \in \mathbf{R}^n$ and $\lambda \in (0, 1)$. If the inequality is reversed, we say that F is *log convex*. We shall sometimes write $F(x) = e^{f(x)}$ and f is concave, but it then is understood that f can take on the value $-\infty$. We say that F is *even* if $F(x) = F(-x)$, $\forall x$.

Two important examples of log concave functions are:

(a) $F(x) = \exp[-(x, Ax)]$, where A is any symmetric real positive-semidefinite quadratic form on $\mathbf{R}^n$.

(b) Let C be any convex set in $\mathbf{R}^n$ and let $\chi_C(x) = 1$ for $x \in C$, $\chi_C(x) = 0$ for $x \notin C$ be the characteristic function of C. Then χ_C is a log concave function. χ_C is even if and only if C is *balanced*, i.e. $x \in C \Rightarrow -x \in C$.

THEOREM 1.1. *Let F be a log concave function on $\mathbf{R}^{m+n}$ and $F: (x, y) \mapsto F(x, y)$ for $x \in \mathbf{R}^m$, $y \in \mathbf{R}^n$. Then $G(x) \equiv \int_{\mathbf{R}^n} F(x, y) \, dy$ is a log concave function on $\mathbf{R}^m$.*

We have four different proofs of this theorem, one of which is the following.

Proof. It is sufficient to prove the theorem when $m = n = 1$; the general case follows by Fubini's theorem and induction. Choose two points x and x' such that $G(x) \neq 0$ and $G(x') \neq 0$. We may assume that

$$\sup_y \{F(x, y)\} = \sup_y \{F(x', y)\},$$

for otherwise we can replace $F(x, y)$ by $e^{bx} F(x, y)$ with b suitably chosen. For each $z \geq 0$, define

$$C(z) = \{(x, y) | F(x, y) \geq z\} \subset \mathbf{R}^2,$$

$C(x, y) = \{y | F(x, y) \geq z\} \subset \mathbf{R}$ and $g(x, z) = \text{meas}\{C(x, z)\}$.
Then
 (i) $C(z)$ is convex and thus $C(x, z)$ is an interval;
 (ii) $G(x) = \int_0^\infty g(x, z) \, dz$;
 (iii) for all $0 \leq \lambda \leq 1$, $g(\lambda x + (1 - \lambda)x', z) \geq \lambda g(x, z) + (1 - \lambda) g(x', z)$.
This last fact follows easily from the convexity of $C(z)$; it is also the Brunn–Minkowski theorem which, in one dimension, is trivial. Thus

$$G(\lambda x + (1 - \lambda)x') \geq \lambda G(x) + (1 - \lambda) G(x') \geq G(x)^\lambda G(x')^{1-\lambda}.$$

Q.E.D.

Theorem 1.1 should not be confused with the following theorem, which is much simpler and which follows directly from Hoelder's inequality.

THEOREM 1.2. *Let $F: \mathbf{R}^{m+n} \to \mathbf{R}$ and, for $x \in \mathbf{R}^m$, $y \in \mathbf{R}^n$, let $F(x, y)$ be log convex in x for each fixed y. Then $G(x) \equiv \int_{\mathbf{R}^n} F(x, y) \, dy$ is log convex on $\mathbf{R}^m$.*

An immediate consequence of Theorem 1.1 is the following.

THEOREM 1.3. *The convolution of two log concave functions on $\mathbf{R}^n$ is log concave.*

Proof. $H(x) = \int_{\mathbf{R}^n} F(x - y) G(y) \, dy$ is log concave since $F(x - y) G(y)$ is jointly log concave in $(x, y) \in \mathbf{R}^{n+m}$.

Q.E.D.

REMARK. In the case of $\mathbf{R}$, Theorem 1.3 is known [1].

1.3. Application of Theorem 1.1 to Gaussian measures

A Gaussian measure on $\mathbf{R}^n$ is given by an (unnormalized) density function $W(x) = \exp[-(x, Ax)/2]$, $A > 0$. The expectation value of a real-valued function H, on $\mathbf{R}^n$, is given by

$$\langle H \rangle_0 = \frac{\int H(x) W(x) \, dx}{\int W(x) \, dx}.$$

Now suppose that $W(x)$ is replaced by $W_F(x) = W(x)F(x)$, where F is a log concave function. With respect to the new weight we define $\langle H \rangle_F$ as above. How does $\langle H \rangle_F$ compare with $\langle H \rangle_0$?

THEOREM 1.4. *The covariance matrix M^F, whose elements are $M_{ij}^F = \langle x_i x_j \rangle_F - \langle x_i \rangle_F \langle x_j \rangle_F$, satisfies*

$$M^F \leq M^0 = A^{-1}$$

in the sense of forms, i.e. $M^0 - M^F$ is positive-semidefinite.

Proof. Consider the function T on $\mathbf{R}^{n+m}$ defined by

$$T(x, y) = W(x)F(x) \exp[-(y, A^{-1}y)/2 + (y, x)]$$
$$= W(x - A^{-1}y)F(x).$$

Then T is log concave and, by Theorem 1.1, $U(y) = \int dx\, T(x, y)$ is log concave on $\mathbf{R}^n$. Thus, the matrix $\partial^2 \ln U(y)/\partial y_i \partial y_j|_{y=0} = M^F - M^0$ is negative-semidefinite.

Q.E.D.

THEOREM 1.5. *If, in the above, we replace F by a log convex function then $M^F \geq M^0 = A^{-1}$.*

Proof. Write $U(y) = \int_{\mathbf{R}^n} W(x) F(x + A^{-1}y)\, dx$ and use Theorem 1.2.

Q.E.D.

As an application of Theorem 1.4, consider an Ising model with Boltzmann factor $B(\sigma) = \exp[\frac{1}{2}(\sigma, A\sigma)]$, $\sigma_i = \pm 1$, $i = 1, \ldots, n$. By adding an unimportant multiple of the identity to A, we can always assume $A > 0$. Since

$$B(\sigma) = (2\pi)^{-n/2} [\det A]^{-\frac{1}{2}} \int_{\mathbf{R}^n} \exp[-(x, A^{-1}x)/2 + (x, \sigma)]\, dx$$

we find that the covariance matrix of the σs, $N_{ij} \equiv \langle \sigma_i \sigma_j \rangle_B$, is simply related to the covariance matrix M^F, introduced above (with A replaced by A^{-1}), by

$$N = A^{-1} M^F A^{-1} - A^{-1},$$

where

$$F(x) = \sum_\sigma e^{(x,\sigma)} = \prod_{i=1}^n 2 \cosh x_i.$$

Now $F(x)$ is log convex, so Theorem 1.5 states that $M^F \geq A$, which implies that $N \geq 0$—hardly an interesting result. Note, however, that

$$G(x) \equiv F(x) \exp\left(-\frac{1}{2} \sum_{i=1}^n x_i^2\right)$$

is log concave. Therefore, provided $A^{-1} > I$ (equivalently, $A < I$) we can write

$$\exp[-(x, A^{-1}x)/2]F(x) = \exp[-(x, (A^{-1}-I)x)/2]G(x)$$

and Theorem 1.4 states that $M^F \leq (A^{-1}-I)^{-1}$ and $N \leq (I-A)^{-1}$. In the physical situation, A is a matrix whose eigenvalues are of $O(1)$ independent of n and $A < I$ occurs for sufficiently high temperature, independently of n. Hence, for high temperature, the eigenvalues of N are $O(1)$; this means there is no long-range order. Although previously there existed elementwise bounds on N for special choices of A ([2] and [3]: inequalities), our result is the first case of a quadratic form inequality on N.

We now quote an assortment of theorems, to indicate some of the directions in which Theorem 1.4 can be generalized.

THEOREM 1.6. *Consider the weight $W_F(x) = W(x)F(x)$ with F log concave, as in Theorem 1.4, and let F be even. Let L be any symmetric, real, n-square matrix. Then*

$$\langle (x, Lx)^2 \rangle_F - \langle (x, Lx) \rangle_F^2 \leq 2\langle (x, LA^{-1}Lx) \rangle_F. \quad (1.1)$$

Proof. We consider the case in which $A = I$; the general case can be handled by the change of variables $x \to A^{-\frac{1}{2}}x$. Let $Z = \int_{\mathbf{R}^n} e^{-(x,x)/2} F(x) \, dx$. Then

$$2\Delta \equiv 2\langle (x, Lx)^2 \rangle_F - 2\langle (x, Lx) \rangle_F^2$$

$$= Z^{-2} \int_{\mathbf{R}^n} \int_{\mathbf{R}^n} e^{-[(x,x)+(y,y)]/2} F(x)F(y)[(x, Lx)-(y, Ly)]^2 \, dx \, dy$$

$$= Z^{-2} \int_{\mathbf{R}^n} \int_{\mathbf{R}^n} e^{-[(u,u)+(v,v)]/2} F[2^{-\frac{1}{2}}(u+v)] F(2^{-\frac{1}{2}}(u-v)) 4(u, Lv)^2 \, du \, dv$$

after the change of variables $x = 2^{-\frac{1}{2}}(u+v)$, $y = 2^{-\frac{1}{2}}(u-v)$. Now do the v integration and recall that $\{\langle v_i, v_j \rangle\}_{i,j=1}^n \leq I$ for each u, by Theorem 1.4. Thus, $2\Delta \leq 4\langle (u, L^2u) \rangle$. Returning to the original x, y variables, one notes that $2(u, L^2u) = (x, L^2x) + (y, L^2y) + 2(x, L^2y)$.

Finally $\langle (x, L^2x) \rangle = \langle (y, L^2y) \rangle = \langle (x, L^2x) \rangle_F$ and $\langle x_i y_j \rangle = 0$.

Q.E.D.

REMARKS. (i) If F is log convex, the inequality in Theorem 1.6 is reversed.

(ii) The significance of Theorem 1.6 is that if L and A are of the order of I, the left side of (1.1) is the difference of two terms of $O(n^2)$, while the right side is $O(n)$. Choosing $L = A$, the left side of (1.1) is like n times a specific heat, while the right side is like n times an internal energy—to use the language of statistical mechanics. Usually, it is difficult to obtain an upper bound on a specific heat.

COROLLARY 1.7. *Let A and L be symmetric, n-square matrices with A non-singular, let F be even and log concave and let λ be real. Then*

$$Z(\lambda) \equiv \int_{\mathbf{R}^n} \exp[-(x, A\,e^{\lambda L}Ax)]F(x)\,dx$$

is log concave in λ.

Proof. Compute $d^2 \ln Z/d\lambda^2$ and compare with Theorem 1.6.

Q.E.D.

THEOREM 1.8. *Let $W_F(x) = e^{-x^2/2}F(x)$ be a weight in $\mathbf{R}$ with F log concave. Define $\langle \,.\, \rangle_F$ and $\langle \,.\, \rangle_0$ as before. Then*

$$\langle |x - \langle x \rangle_F|^\alpha \rangle_F \leq \langle |x - \langle x \rangle_0|^\alpha \rangle_0$$

for $\alpha \geq 1$.

The proof of Theorem 1.8 is lengthy and will not be given here. The theorem says that multiplying a Gaussian weight on $\mathbf{R}$ by a log concave function may, if the function is not even, shift the mean, but all moments, higher than the first, with respect to the new mean are decreased.

We present next a theorem which will play an important role in the next section.

THEOREM 1.9. *Let A be a real positive-definite $(n+m)$-square matrix partitioned as $A = \begin{bmatrix} \alpha & \beta \\ \beta^T & \gamma \end{bmatrix}$, where α is n-square, γ is m-square, β is $n \times m$, and T means transpose. Let F be a log concave function on $\mathbf{R}^{n+m}$ and form the unnormalized weight on $\mathbf{R}^{n+m}$: $W_F(x) = W(x)F(x)$, $W(x) = \exp[-(x, Ax)/2]$. Denoting, as before, a point $x \in \mathbf{R}^{m+n}$ as $x = (y, z)$, $y \in \mathbf{R}^n$, $z \in \mathbf{R}^m$, define the unnormalized weight V on $\mathbf{R}^n$ by*

$$V(y) = \int_{\mathbf{R}^m} W_F(y, z)\,dz.$$

If we define $G: \mathbf{R}^n \to \mathbf{R}$ by

$$V(y) = \exp[-(y, By)/2]G(y),$$

with $B = \alpha - \beta\gamma^{-1}\beta^T > 0$, then G is log concave.

Proof. Note that the $(n+m)$-square matrix $C \equiv A - \begin{bmatrix} B & 0 \\ 0 & 0 \end{bmatrix}$ is positive-semidefinite, since A is positive-definite. Hence $U_F(x) = \exp[-(x, Cx)/2]F(x)$ is log concave on $\mathbf{R}^{n+m}$. Since

$$V(y) = \exp[-(y, By)/2]\int_{\mathbf{R}^m} U_F(y, z)\,dz,$$

Theorem 1.9 follows from Theorem 1.1.

Q.E.D.

REMARKS. (i) *Mutatis mutandis*, if F is replaced by a log convex function, then G is log convex on $\mathbf{R}^n$.

(ii) If $F(x)$ is a constant, then $G(y)$ is also a constant. Thus, Theorem 1.9 states that if one does a partial integration over a Gaussian weight times a log concave function, the result is the Gaussian weight one would have obtained without the log concave multiplier times a new log concave function.

To pursue the ideas of Theorem 1.9 a bit further, let us formulate the Brunn–Minkowski theorem for Gaussian measures. We recall the classical Brunn–Minkowski theorem [4].

THEOREM 1.10. *Let C_0, C_1 be non-empty convex sets in $\mathbf{R}^n$, and let*

$$C_\lambda = \lambda C_1 + (1-\lambda)C_0, \qquad 0 \leq \lambda \leq 1.$$

Denote by $|C|$ the n-dimensional Lebesgue measure of C. Then

$$|C_\lambda|^{1/n} \geq \lambda |C_1|^{1/n} + (1-\lambda)|C_0|^{1/n}.$$

REMARK. If $C_0 = \{0\}$, then $C_\lambda = \lambda C_1$.

In the case of Gaussian measures we have the following.

THEOREM 1.11. *Let C_0, C_1 and C_λ be as in Theorem 1.10, and let A be a real, positive-definite, n-square matrix. Let*

$$\mu_G(C) = \int_C \exp[-(x, Ax)/2]\, dx.$$

Then

$$\mu_G(C_\lambda) \geq \mu_G(C_1)^\lambda \mu_G(C_0)^{1-\lambda}.$$

Proof. Define the convex set

$$D = \{(\lambda, x) | 0 \leq \lambda \leq 1, x \in C_\lambda\} \subset \mathbf{R} \times \mathbf{R}^n.$$

Then

$$\mu_G(C_\lambda) = \int_{\mathbf{R}} \chi_D(\lambda, x) \exp[-(x, Ax)/2]\, dx.$$

Since the integrand is log concave in (λ, x), $\mu_G(C_\lambda)$ is log concave by Theorem 1.1.

Q.E.D.

As a corollary to Theorem 1.11 we quote a theorem of L. Gross [5]. The Gaussian measure μ_n on $\mathbf{R}^n$ defines the measure of a Borel set $B \subset \mathbf{R}^n$ to be

$$\mu_n(B) = (2\pi)^{-n/2} \int_B \exp[-(x,x)/2]\,dx.$$

THEOREM 1.12. (L. Gross) *Let C be a convex, balanced set in $\mathbf{R}^n \times \mathbf{R}^m$, let D be the intersection of C with $\mathbf{R}^m$, and let E be the projection of C on $\mathbf{R}^n$. Then*

$$\mu_{n+m}(C) \leq \mu_n(E)\mu_m(D).$$

Proof. Let C_x be the intersection of C with the plane parallel to $\mathbf{R}^m$ through $x \in \mathbf{R}^n$; in particular, $C_0 = D$. By the symmetry of C and Theorem 1.11, $x \to \mu_m(C_x)$ is log concave and even on $\mathbf{R}^n$, and hence $\mu_m(C_x)$ is maximal for $x = 0$. Thus

$$\mu_{n+m}(C) = \int_E \mu_m(C_x)\,d\mu_n(x) \leq \mu_m(D)\int_E d\mu_n(x) = \mu_n(E)\mu_m(D).$$

Q.E.D.

Let us return to the Brunn-Minkowski theorem (1.11) for Gaussians. By passing to the limit $n \to \infty$, the same theorem obviously remains true for infinite-dimensional Gaussian measures, for example, the Wiener measure. In that case we should deal with measurable, convex sets of Wiener paths. We shall consider here particular convex sets of paths, namely those passing through a convex set $C_\lambda \subset \mathbf{R}^n$ for all t.

With C_λ, $0 \leq \lambda \leq 1$, defined as in Theorem 1.10, consider the fundamental solution $G_\lambda(x, y; t)$, $t \geq 0$, of the diffusion equation with potential V, in $\mathbf{R}^n$, defined by

$$\left(\frac{\partial}{\partial t} - \frac{1}{2}\Delta_x + V(x)\right)G_\lambda(x, y; t) = 0;$$

$$G_\lambda(x, y; 0) = \delta(x-y), \qquad x, y \in C_\lambda;$$

$$G_\lambda(x, y; t) = 0, \qquad x \in \partial C_\lambda, \qquad \forall y, t;$$

$$\equiv 0, \qquad x \notin C_\lambda \text{ or } y \notin C_\lambda.$$

THEOREM 1.13. *Let $V(x)$ be a convex function. Then $G_\lambda(x, y; t)$ is log concave in $(x, y, \lambda) \in \mathbf{R}^n \times \mathbf{R}^n \times [0, 1]$.*

Proof. Use the Trotter product formula with $x_0 = x$, $x_N = y$:

$$G_\lambda(x, y; t) = \lim_{N \to \infty} (2\pi t/N)^{-nN/2} \int_{\mathbf{R}^{n(N-1)}} dx_1 \ldots dx_{N-1} \times$$

$$\times \prod_{j=1}^N \left\{\exp\left[-\frac{N}{2t}(x_j - x_{j-1})^2 - \frac{t}{N}V(x_j)\right]\chi_{C_\lambda}(x_j)\right\}.$$

The integrand is log concave in $(x, x_1, \ldots, x_{N-1}, y, \lambda)$. Finally the pointwise limit of a sequence of log concave functions is log concave.

Q.E.D.

COROLLARY 1.14. *In addition to the hypotheses of Theorem 1.13, either let C_0 and C_1 be compact or let $\exp(-tV)$ be in $L^1(\mathbf{R}^n)$, $\forall t > 0$. Define*

$$Z_\lambda(t) = \int_{C_\lambda} G_\lambda(x, x; t) \, dx = \mathrm{tr} \, e^{-tH},$$

with $H = -\tfrac{1}{2}\Delta + V$. Then $Z_\lambda(t) < \infty$ and $Z_\lambda(t)$ is log concave in λ.

Proof. That $Z_\lambda(t)$ is finite is a standard result and can be proved from the Trotter product formula above using Hoelder's inequality. The log concavity of $Z_\lambda(t)$ follows from Theorems 1.1 and 1.13.

Q.E.D.

COROLLARY 1.15. *Let $V(x)$ and C_λ be as in Corollary 1.14. Let $\varepsilon_0(\lambda)$ be the lowest eigenvalue of the equation*

$$[-\tfrac{1}{2}\Delta + V(x)]f(x) = \varepsilon_0(\lambda)f(x),$$

with $f(x) = 0$ for $x \in \partial C_\lambda$. Then $\varepsilon_0(\lambda)$ is a convex function of $\lambda \in [0, 1]$.

Proof. Since e^{-tH} is trace class, $Z_\lambda = \sum_{j=0}^{\infty} \exp[-t\varepsilon_j(\lambda)]$, $\varepsilon_{j+1}(\lambda) \geq \varepsilon_j(\lambda)$ and each $\varepsilon_j(\lambda)$ has finite multiplicity. Then

$$\varepsilon_0(\lambda) = -\lim_{t \to \infty} t^{-1} \ln Z_\lambda(t),$$

and, since the pointwise limit of a sequence of convex functions is convex, Corollary 1.15 is proved.

Q.E.D.

1.4. The one-dimensional plasma

In this section we apply the previous theorems to an old problem in physics, namely, to the one-dimensional, one-component plasma in a neutralizing background. We shall consider both the classical and quantum-mechanical cases. The latter requires the introduction of the Wiener integral, and thus provides another example of the application of our theorems to functional integrals. The object of our investigations is to show that long-range order exists for sufficiently large coupling constant, i.e. that the one-particle distribution function is a non-constant periodic function. The occurrence of this phenomenon was first predicted by Wigner [6].

Let $x = (x_{-n}, \ldots, x_n)$ be the coordinates of $(2n+1)$ one-dimensional particles, each having a negative charge of one unit. The one-dimensional Coulomb potential between two unit charges separated by a distance $|x|$ is $-|x|$. Then the total potential energy of $(2n+1)$ particles in a 'box' $[-L, L]$ with a fixed uniform positive charge background of density ρ is

$$\phi(x) = -\sum_{-n \leq i < j \leq n} |x_i - x_j| + \rho \sum_{i=-n}^{n} \int_{-L}^{L} |x_i - x| \, dx - \frac{\rho^2}{2} \int_{-L}^{L} \int_{-L}^{L} |x - y| \, dx \, dy. \tag{1.2}$$

We shall further assume that the total charge is zero, i.e.

$$2L\rho = 2n + 1.$$

Since ϕ is symmetric in the x_i, it is sufficient to consider the convex domain

$$C = \{x \mid -L \leq x_{-n} \leq x_{-n+1} \ldots \leq x_n \leq L\}. \tag{1.3}$$

In C,

$$\phi(x) = \rho \sum_{j=-n}^{n} \left(x_j - \frac{j}{\rho}\right)^2, \tag{1.4}$$

where a constant term in the potential has been neglected. Our methods are capable of handling the domain C as it stands, but then the function we wish to calculate, $\rho(x)$, will not be strictly periodic, except in the thermodynamic limit $n \to \infty$, $\rho = $ constant. To circumvent this difficulty we extend C to the larger convex domain

$$D = \{x \mid x_{-n} \leq x_{-n+1} \ldots \leq x_n \leq x_{-n} + 2L\}. \tag{1.5}$$

The domain D no longer confines the particles to the box, and we shall cheat a little by supposing that the expression (1.4) for ϕ is valid in all of D. The original walls of the box are still visible in ϕ.

Remark then that the domain D and the potential ϕ are invariant under the linear transformations R(reflection) and T(translation), defined by

$$(Rx)_j = -x_{-j} \tag{1.6}$$

$$(Tx)_j = x_{j+1} - \frac{1}{\rho}, \qquad -n \leq j \leq n-1;$$

$$(Tx)_n = x_{-n} + \frac{2n}{\rho}. \tag{1.7}$$

1.4.1. The classical case

The Gibbs distribution function of the jth particle is defined by (β is the reciprocal temperature)

$$\rho_j^{(n)}(x) = \frac{\int_D \delta(x - x_j) \exp[-\beta \phi(x)] \, dx}{\int_D \exp[-\beta \phi(x)] \, dx}.$$

The symmetry properties (1.6) and (1.7) imply that

$$\rho_0^{(n)}(x) = \rho_0^{(n)}(-x); \qquad (1.8)$$

$$\rho_j^{(n)}(x) = \rho_0^{(n)}\left(x - \frac{j}{\rho}\right), \quad -n \leq j \leq n. \qquad (1.9)$$

Since D is a convex domain in $\mathbf{R}^{2n+1}$, direct application of Theorem 1.9 gives that

$$\rho_0^{(n)}(x) = \exp(-\beta\rho x^2) F^{(n)}(x), \qquad (1.10)$$

where $F^{(n)}$ is log concave; by eqn (1.8), $F^{(n)}(x)$ is also even.

We shall not go into the details of the existence of the limiting distribution functions

$$\rho_j(x) \equiv \lim_{n \to \infty} \rho_j^{(n)}(x)$$

in the thermodynamic limit ($n \to \infty$, $L \to \infty$, $2n+1 = 2L\rho$). Obviously, properties (1.8)–(1.10) remain true in the limit. It is also fairly clear, that the use of domain C instead of domain D would give the same distribution functions in the limit.

Thus far we have established part (*i*) of the following theorem:

THEOREM 1.16. (*i*) *The one-particle distribution functions of the classical one-dimensional plasma computed in D satisfy*

$$\rho_0(x) = \rho_0(-x),$$

$$\rho_j(x) = \rho_0\left(x - \frac{j}{\rho}\right),$$

$$\rho_0(x) = \exp[-\beta\rho x^2] F(x),$$

where $F(x)$ is a log concave, even function.

(*ii*)
$$\int_{\mathbf{R}} |x|^\alpha \rho_0(x) \, dx \leq \left(\frac{\pi}{\beta\rho}\right)^{-\frac{1}{2}} \int_{\mathbf{R}} |x|^\alpha \exp[-\beta\rho x^2] \, dx$$

for $\alpha \geq 1$.

(*iii*) *For large values of β/ρ, the total density*

$$\rho(x) = \sum_{j=-\infty}^{\infty} \rho_j(x)$$

is non-trivially periodic.

Proof. (*ii*) Follows from Theorem 1.8, (*iii*) will be proved in § 1.4.3, Theorem 1.18.

REMARKS. (*i*) Theorem 1.16 (*ii*) states that the moments of the one-particle distribution functions are smaller than they would be without the restriction $x_j \leq x_{j+1}$.

(*ii*) The interpretation of Theorem 1.16 (*iii*) is that the plasma is in a crystalline state. The specific position of the crystal is a consequence of the hard walls that were imposed at $\pm L = \pm(n+\frac{1}{2})\rho$. This fact is reflected not only in the domain D (eqn (1.5)) but also in the expressions (1.4) for $\phi(\mathbf{x})$. Hard walls at positions $\pm L + \delta$ would translate the crystal through a distance δ.

(*iii*) The fact that $\rho(x)$ is not a constant has recently been proved by Kunz [7] who, by other methods, showed that to be true for all β, ρ except possibly for a countable number of values of β/ρ.

1.4.2. The quantum-mechanical case

The quantum-mechanical Hamiltonian of the system defined by equation (1.2), with $\hbar^2/m = 1$, is

$$H = -\tfrac{1}{2}\Delta + \phi(\mathbf{x}),$$

where

$$\Delta = \sum_{j=-n}^{n} \frac{\partial^2}{\partial x_j^2}.$$

We consider the case that the particles are spinless fermions. This means that H acts on square integrable, antisymmetric functions. As is well known, an equivalent statement in one dimension is that H acts on square integrable functions defined on $E = \{\mathbf{x} | x_{-n} \leq x_{-n+1} \leq \ldots \leq x_n\}$ which vanish on the boundary of E. The 'box' condition requires that the functions vanish on the boundary of $C \subset E$. As in the classical case, we shall use the larger domain D instead of C.

The distribution function of the jth particle is then

$$\rho_j^{(n)}(x) = \frac{\operatorname{tr}_D e^{-\beta H} \delta_j(x)}{\operatorname{tr}_D e^{-\beta H}},$$

where $\delta_j(x)$ is the operator of multiplication by $\delta(x - x_j)$. Since H and D are invariant under the transformations R and T (eqns (1.6), (1.7)), the distribution functions again have the symmetry properties ((1.8), (1.9)).

To find the analogue of eqn (1.10), use the Trotter product formula for $\exp(-\beta H)$, which gives, with $\mathbf{x}^0 = \mathbf{x}^N$,

$$\operatorname{Tr}_D e^{-\beta H} \delta_0(x) = \lim_{N \to \infty} \left(\frac{2\pi\beta}{N}\right)^{-(2n+1)N/2} \int_{D^N} d\mathbf{x}^1 \ldots d\mathbf{x}^N \times$$

$$\times \prod_{k=1}^{N} \exp\left[-\frac{N}{2\beta} \sum_{j=-n}^{n} (x_j^k - x_j^{k-1})^2 - \frac{\beta\rho}{N} \sum_{j=-n}^{n} \left(x_j^k - \frac{j}{\rho}\right)^2\right] \delta(x_0^1 - x). \quad (1.11)$$

Since D^N is convex in $\mathbf{R}^{(2n+1)N}$, we can apply Theorem 1.9 to conclude that

$$\rho_0^{(n)}(x) = \exp(-\gamma x^2) H^{(n)}(x), \qquad (1.12)$$

where $H^{(n)}$ is log concave, and where $\exp(-\gamma x^2)$ is, up to multiplication by an x-independent constant, what eqn (1.11) would give if D were replaced by $\mathbf{R}^{2n+1}$. But in that case the integrations separate into $(2n+1)$ independent integrations over $\mathbf{R}^N$. Therefore, $\exp(-\gamma x^2)$ is proportional to $G(x, x; \beta)$, where $G(x, y; t)$ is the fundamental solution (Green's function) of the differential equation, for $t > 0$,

$$\left(\frac{\partial}{\partial t} - \frac{1}{2}\frac{\partial^2}{\partial x^2} + \rho x^2\right) G(x, y; t) = 0;$$

$$G(x, y; 0) = \delta(x - y).$$

Using the well-known expression [8] for G, we obtain

$$\gamma = (2\rho)^{\frac{1}{2}} \tanh \beta(\rho/2)^{\frac{1}{2}}. \qquad (1.13)$$

The analogue of Theorem 1.16 is now immediate.

THEOREM 1.17: *Theorem 1.16 is correct for the quantum-mechanical plasma of spinless fermions except that in part (ii) $\beta\rho$ is replaced by γ (eqn (1.13)) and in part (iii) β/ρ is replaced by γ/ρ^2.*

Remarks (i) and (ii) after Theorem 1.16 also apply here.

We turn next to the demonstration that parts (iii) of Theorems 1.16 and 1.17 follow from parts (i) of those theorems.

1.4.3. Can modified theta functions be constant?

Let $f(x) = \exp(-\lambda x^2) F(x)$ with $F(x)$ even and log concave and $\lambda > 0$. Consider

$$\rho(x) = \sum_{j=-\infty}^{\infty} f(x - j). \qquad (1.14)$$

The question to which we address ourselves here is whether or not F can be chosen so that $\rho(x)$ is constant. The answer, surprisingly, depends on λ. As Theorem 1.18 shows, $\rho(x)$ cannot be constant when λ is large, and thus parts (iii) of Theorems 1.16 and 1.17 are proved.

Define the Fourier transform by

$$\hat{f}(k) = \int_{\mathbf{R}} e^{2\pi i k x} f(x) \, dx. \qquad (1.15)$$

Then, by the Poisson summation formula,

$$\rho(x) = \sum_{j=-\infty}^{\infty} \hat{f}(j) e^{-2\pi i j x}. \qquad (1.16)$$

Therefore $\rho(x)$ is constant iff $\hat{f}(j) = 0$ for $j = \pm 1, \pm 2, \ldots$.

THEOREM 1.18. *Let $\rho(x)$ be defined as in eqn (1.14). Then there exists a λ_0, $0 < \lambda_0 < \infty$, such that*
  (a) *For all $\lambda > \lambda_0$ and for all F even and log concave, $\rho(x)$ is not constant;*
  (b) *For all $\lambda < \lambda_0$, $\lambda > 0$ there exists an even, log concave F such that $\rho(x) = $ constant.*

Proof. (i) Existence of λ_0: If, for some λ, there is an $F(x)$ that leads to a constant $\rho(x)$, then, for $\mu < \lambda$, the log concave function $F(x) \exp[(\mu - \lambda)x^2]$ gives the same constant $\rho(x)$.
 (ii) $\lambda_0 < \infty$: Normalize to $F(0) = 1$. Then $\rho(0) \geq 1$, and

$$\rho(\tfrac{1}{2}) \leq \sum_{j=-\infty}^{\infty} \exp[-\lambda(j - \tfrac{1}{2})^2] \leq 2 e^{-\lambda/4} \sum_{j=0}^{\infty} e^{-2\lambda j} = 2 e^{-\lambda/4}(1 - e^{-2\lambda})^{-1}.$$

This gives the simple estimation $\lambda_0 < 3$.
 (iii) $\lambda_0 > 0$: We indicate how to construct an example of constant ρ for λ sufficiently small. Choose a non-constant, even, log concave function G, and normalize it so that $g(x) \equiv \exp(-\lambda x^2) G(x)$ satisfies

$$\int_{\mathbf{R}} g(x) \, dx = \hat{g}(0) = 1.$$

Define

$$\hat{f}(k) = \prod_{j=1}^{\infty} \hat{g}(k/j), \qquad (1.17)$$

which is the Fourier transform of the convolution

$$f(x) = \prod_{j=1}^{\infty} {}^*j \exp(-\lambda j^2 x^2) G(jx). \qquad (1.18)$$

The infinite product (1.17) is defined and $\hat{g}(k) > 0$ in a neighbourhood of $k = 0$, since

$$1 > \hat{g}(k/j) \approx 1 + \tfrac{1}{2}(k/j)^2 \hat{g}''(0) > 0 \quad \text{for} \quad |k/j| \ll 1.$$

Equation (1.18) then follows from the Lebesgue dominated convergence theorem, and $f \neq 0$. Now Theorem 1.9 applied to eqn (1.18) gives

$$f(x) = \exp(-\alpha \lambda x^2) F(x),$$

where $F(x)$ is log concave and even and $\alpha = (\sum_{j=1}^{\infty} j^{-2})^{-1} = 6\pi^{-2}$.
 It is now sufficient to determine λ and G such that $\hat{g}(\pm 1) = 0$; then, by eqn (1.17),

$$\hat{f}(j) = 0 \quad \text{for all integers} \quad j \neq 0,$$

and we are done. Take $G(x) = [1 + \sqrt{(\lambda/\pi)}]\chi(x)$, where χ is the characteristic function of $[-\frac{3}{4}, \frac{3}{4}]$. Then

$$\lim_{\lambda \to 0} \hat{g}(k) = (\pi k)^{-1} \sin \tfrac{3}{2}\pi k; \lim_{\lambda \to \infty} \hat{g}(k) = 1.$$

Therefore λ can be chosen such that $\hat{g}(\pm 1) = 0$.

Q.E.D.

Acknowledgements

This work has been partially supported by U.S. National Science Foundation Grants GP-31674X and GP-16147A #1.

References

1. SCHOENBERG, I. J. On Pólya frequency functions I: The totally positive functions and their Laplace transforms. *J. Anal. math.* **1**, 331–74 (1951).
2. GRIFFITHS, R. B. Correlations in Ising ferromagnets, I, II, III. *J. Math. Phys.* **8**, 478–83, 484–9 (1967); *Commun. Math. Phys.* **6**, 121–7 (1967); KELLY, D. and SHERMAN, S. General Griffiths inequalities on correlations in Ising ferromagnets. *J. Math. Phys.* **9**, 466–84 (1968).
3. FORTUIN, C. M., KASTELEYN, P. W., and GINIBRE, J. Correlation functions on some partially ordered sets. *Commun. Math. Phys.* **22**, 89–103 (1971).
4. BONNESEN, T. and FENCHEL, W. *Theorie der Konvexen Koerper.* Chelsea, New York (1948).
5. GROSS, L. Measurable functions on Hilbert space. *Trans. Am. math. Soc.* **105**, 372–90 (1962); see also DUDLEY, R. M., FELDMAN, J., and LE CAM, L. On seminorms and probabilities, and abstract Wiener spaces. *Ann. Math.* **93**, Ser. 2, 390–408 (1971).
6. WIGNER, E. P. Effects of the electron interaction on the energy levels of electrons in metals. *Trans. Faraday Soc.* **34**, 678–85 (1938).
7. KUNZ, H. Equilibrium properties of the one-dimensional classical electron gas. Preprint E.P.F.L., Lausanne (1974).
8. MERZBACHER, E. *Quantum mechanics*, Chapter 8. Wiley, New York (1961).

Note

Since this work was completed we have found that Theorem 1.1 has been proved independently by A. Prekopa, *Acta Math. Szeged* **32**, 301–15 (1971), and Y. Rinott, Thesis, Weizmann Institute, Rehovoth, Israel (Nov. 1973).

2. Asymptotic evaluation of certain Wiener integrals for large time

M. D. DONSKER AND S. R. S. VARADHAN

2.1. Introduction

LET $V(y) \geq 0$ and continuous on $-\infty < y < \infty$. If $V(y) \to +\infty$ as $y \to \pm\infty$, the eigenvalue problem

$$\tfrac{1}{2}\psi''(y) - V(y)\psi(y) = -\lambda \psi(y) \tag{2.1}$$

has a discrete spectrum, and an old result of Kac [1] is that, for the least eigenvalue λ_1,

$$\lim_{t \to \infty} \frac{1}{t} \ln E_x\left\{\exp\left[-\int_0^t V(z(\tau))\,d\tau\right]\right\} = -\lambda_1 \tag{2.2}$$

where $E_x\{\ \}$ denotes expectation on the Brownian-motion process $\{z_\tau, 0 \leq \tau < \infty\}$ with $z(0) = x$.

The formal idea of his proof was that if, for any real y, we let $u(t, x, y)$ be the solution of

$$\frac{\partial u}{\partial t} = \frac{1}{2}\frac{\partial^2 u}{\partial x^2} - V(x)u,$$

$$u(x, 0) = \delta(x - y),$$

then, on the one hand, by the Feynman–Kac formula,

$$u(t, x, y) = E_x\left\{\exp\left[-\int_0^t V(z(\tau))\,d\tau\right]\delta(z(t) - y)\right\}$$

and, on the other hand,

$$u(t, x, y) = \sum_{j=1}^{\infty} \exp(-\lambda_j t)\psi_j(x)\psi_j(y)$$

where $\{\lambda_j\}$ and $\{\psi_j\}$ are respectively the eigenvalues and normalized eigenfunctions of (2.1). Thus, integrating on y

$$E_x\left\{\exp\left[-\int_0^t V(z(\tau))\,d\tau\right]\right\} = \sum_{j=1}^{\infty} \exp(-\lambda_j t)\psi_j(x)\int_{-\infty}^{\infty}\psi_j(y)\,dy$$

which gives (2.2).

A long-standing question has been the relation between (2.2) and the fact that we also know

$$\lambda_1 = \inf\left\{\int_{-\infty}^{\infty} V(y)\psi^2(y)\,dy + \tfrac{1}{2}\int_{-\infty}^{\infty}[\psi'(y)]^2\,dy\right\} \tag{2.3}$$

where the infimum is taken over all $\psi \in L^2$ such that $\int_{-\infty}^{\infty} \psi^2(y)\,dy = 1$. It is reasonable to expect that an expression like the right side of (2.3) should come from a direct asymptotic evaluation of $E_x\{\exp[-\int_0^t V(z(\tau))\,d\tau]\}$. Moreover, if such a direct connection can be made, one should be able to find the asymptotic behaviour of Brownian-motion expectations of more general functionals—including those for which there is no associated differential equation at all. Such a theorem is proved in this chapter.

Let $\mathscr{F}$ be the space of probability distribution functions $F(y)$, $-\infty < y < \infty$, with the Lévy metric:

$$\rho(F_1, F_2) = \inf_{h>0} [F_1(y-h) - h \leq F_2(y) \leq F_1(y+h) + h \text{ for all } y].$$

Let $\Phi(F)$ be a function on $\mathscr{F}$ satisfying:
(1) $0 \leq \Phi(F) \leq \infty$.
(2) $\Phi(F)$ is lower semicontinuous on $\mathscr{F}$.
(3) $\{F \in \mathscr{F} : \Phi(F) \leq K\}$ is a compact subset of $\mathscr{F}$ for every finite $K > 0$.
(4) If $F_n \Rightarrow F$ and the support of F_n is contained in an interval $[a, b]$ for all n, then $\Phi(F_n) \to \Phi(F)$.
(5) Let $F \in \mathscr{F}$ be such that $\Phi(F) < \infty$ and let $\{g_n(y)\}$ be a sequence of continuous functions such that $0 \leq g_n(y) \leq 1$, $-\infty < y < \infty$, with $g_n(y) = 1$ for $|y| \leq n$. If we define $F_n(y) = \int_{-\infty}^{y} g_n(\xi)\,dF(\xi) \bigg/ \int_{-\infty}^{\infty} g_n(\xi)\,dF(\xi)$,

then $\Phi(F_n) \to \Phi(F)$.

If $F \in \mathscr{F}$ has a density function f, we will sometimes for convenience write $\Phi(f)$ for $\Phi(F)$.

Let Ω_x be the space of continuous functions $z(\tau)$, $0 \leq \tau < \infty$, with $z(0) = x$. For $t > 0$ and $\omega \in \Omega_x$, we define

$$L(t, \omega, y) = \frac{1}{t} \lambda\{s : 0 \leq s \leq t, z(s) \leq y\},$$

where $\lambda\{\ \}$ denotes Lebesgue measure, i.e. $L(t, \omega, y)$ is the proportion of time on $[0, t]$ that a particular path $\omega = z(\cdot)$ is less than or equal to y. We note that for each ω and each $t > 0$, $L(t, \omega, \cdot) \in \mathscr{F}$. Since each ω is bounded on $[0, t]$, $L(t, \omega, y)$ is a distribution function whose support is $[\inf_{0 \leq \tau \leq t} z(\tau), \sup_{0 \leq \tau \leq t} z(\tau)]$. $L(t, \omega, y)$ has a density function, $l(t, \omega, y)$, often called the 'local time', and conveniently expressed as

$$\frac{1}{t} \int_0^t \delta(z(\tau) - y)\,d\tau.$$

Let f denote the space of probability density functions f which are continuously differentiable on $-\infty < y < \infty$ and either $f(y) > 0$ for all $-\infty < y < \infty$ or f has compact support and is strictly positive on the interior of that support. The main result of this chapter is the following.

THEOREM 2.1. *Let $\Phi(F)$ satisfy hypotheses (1)–(5) above. Then*

$$\lim_{t\to\infty}\frac{1}{t}\ln E_x\{e^{-t\Phi[L(t,\omega,\cdot)]}\} = -\inf_{f\in\mathscr{f}}\left\{\frac{1}{8}\int_{-\infty}^{\infty}\frac{[f'(y)]^2}{f(y)}\,dy + \Phi(f)\right\}, \qquad (2.4)$$

where $E_x\{\ \}$ denotes integration on Ω_x with respect to Brownian-motion measure.

Let us look at some examples of this theorem. In the particular case, $\Phi(F) = \int_{-\infty}^{\infty} V(y)\,dF(y)$, where $V(y) \geq 0$ and is continuous, and $V(y) \to +\infty$ as $y \to \pm\infty$, it is easy to see that $\Phi(F)$ satisfies hypotheses (1)–(5).† In this case,

$$\Phi(\dot{L}(t,\omega,\cdot)) = \int_{-\infty}^{\infty} V(y)\,dL(t,\omega,y) = \frac{1}{t}\int_0^t V(z(\tau))\,d\tau,$$

and if, for any $f \in \mathscr{f}$, we let $\psi(y) = \sqrt{(f(y))}$, then $\psi \in L^2$ and, indeed, $\int_{-\infty}^{\infty} \psi^2(y)\,dy = 1$. Moreover,

$$\int_{-\infty}^{\infty} V(y)\,dF(y) = \int_{-\infty}^{\infty} V(y)f(y)\,dy = \int_{-\infty}^{\infty} V(y)\psi^2(y)\,dy$$

and

$$\frac{1}{8}\int_{-\infty}^{\infty}\frac{[f'(y)]^2}{f(y)}\,dy = \frac{1}{2}\int_{-\infty}^{\infty}[\psi'(y)]^2\,dy.$$

Thus, in the special case $\Phi(F) = \int_{-\infty}^{\infty} V(y)\,dF(y)$, (2.4) becomes

$$\lim_{t\to\infty}\frac{1}{t}\ln E_x\left\{\exp\left[-\int_0^t V(z(\tau))\,d\tau\right]\right\}$$

$$= -\inf\left\{\int_{-\infty}^{\infty} V(y)\psi^2(y)\,dy + \frac{1}{2}\int_{-\infty}^{\infty}[\psi'(y)]^2\,dy\right\} \qquad (2.5)$$

where the infimum is taken over all $\psi \in L^2$ such that $\int_{-\infty}^{\infty} \psi^2(y)\,dy = 1$. This solves the problem mentioned earlier.

If we let

$$\Phi(F) = \int_{-\infty}^{\infty}\int_{-\infty}^{\infty} \rho(y_1, y_2)\,dF(y_1)\,dF(y_2),$$

where $\rho(y_1, y_2)$ is continuous and $\rho(y_1, y_2) \to +\infty$ if $|y_1|+|y_2| \to \infty$, then $\Phi(F)$ satisfies hypotheses (1)–(5) and (2.4) becomes

$$\lim_{t\to\infty}\frac{1}{t}\ln E_x\left\{\exp\left[-\frac{1}{t}\int_0^t\int_0^t \rho(z(s), z(\tau))\,ds\,d\tau\right]\right\}$$

$$= -\inf_{f\in\mathscr{f}}\left\{\frac{1}{8}\int_{-\infty}^{\infty}\frac{[f'(y)]^2}{f(y)}\,dy + \int_{-\infty}^{\infty}\int_{-\infty}^{\infty} \rho(y_1, y_2)f(y_1)f(y_2)\,dy_1\,dy_2\right\}. \qquad (2.6)$$

† The condition $V(y) \to +\infty$ as $y \to \pm\infty$ is used only for hypothesis (3).

If $\Phi(F) = \exp[\int_{-\infty}^{\infty} y^2 \, dF(y)]$, one obtains readily from (2.4) that

$$\lim_{t \to \infty} \frac{1}{t} \ln E_x\left(\exp\left\{-t \exp\left[\frac{1}{t}\int_0^t z^2(\tau) \, d\tau\right]\right\}\right) = -\frac{\alpha+1}{8\alpha^2},$$

where $\alpha > 0$ is the root of $e^u = 1/8u^2$.

Before we proceed with the proof we shall try to explain the idea behind the proof. By the ergodic theorem, since Brownian motion has no finite invariant measure on the line, as time increases, most of the Brownian paths will spend only a negligible proportion of time in a finite region. However, there will be a few Brownian paths, with very low probability, which hang around too long. Let us fix a nice probability density $f(y)$ and ask the following question. What is the probability that the occupation distribution $L(t, \omega, y)$ of a Brownian path is close to the distribution $F(y)$ with density $f(y)$? The probability will of course go to zero with t: the question is, at what rate? It turns out that, roughly,

$$\text{Prob}\{L(t, \omega, \cdot) \sim F(\cdot)\} \sim \exp\left\{-\frac{t}{8}\int_{-\infty}^{\infty} \frac{[f'(y)]^2}{f(y)} \, dy\right\}.$$

For practical purposes one can view the measure Q_x^t on the space of distribution functions as behaving like

$$Q_x^t \sim \exp\left\{-\frac{t}{8}\int_{-\infty}^{\infty} \frac{[f'(y)]^2}{f(y)} \, dy\right\} df.$$

The theorem is then a Laplace asymptotic formula.

This type of reasoning generalizes to a wide class of Markov processes both with discrete and continuous time and will be dealt with in several forthcoming articles.

2.2. The upper estimate

Let P_x denote Brownian-motion measure on Ω_x. For each x and $t > 0$ we define a probability measure $Q_{x,t}$ on $\mathcal{F}$ by

$$Q_{x,t}(C) = P_x\{\omega : L(t, \omega, \cdot) \in C\}$$

for measurable sets $C \subset \mathcal{F}$, i.e. for which the right side is defined.

Denoting expectation with respect to the measure $Q_{x,t}$ by $E^{Q_{x,t}}\{\ \}$, we have

$$E_x\{e^{-t\Phi[L(t,\omega,\cdot)]}\} = E^{Q_{x,t}}(e^{-t\Phi(F)}). \tag{2.7}$$

In order to get an upper estimate on the left side of (2.7) we first get an upper estimate on $Q_{x,t}(C)$ for appropriate sets C. Let $\mathcal{U}$ be the space of functions $u(y)$, $-\infty < y < \infty$, having two bounded continuous derivatives and for each of which there are two positive numbers α and β such that for all y, $0 < \alpha \leq u(y) \leq \beta < \infty$.

For $u \in \mathcal{U}$, define

$$\psi(x, t) = E_x\left\{u(z(t)) \exp\left[-\int_0^t \left(\frac{u''}{2u}\right)(z(s))\, ds\right]\right\}.$$

From the Feynman–Kac formula we see that $\psi(x, t)$ satisfies

$$\frac{\partial \psi}{\partial t} = \frac{1}{2}\frac{\partial^2 \psi}{\partial x^2} - \frac{u''(x)}{2u(x)}\psi,$$

but, since the solution of this last is clearly $\psi(x, t) = u(x)$, we have

$$E_x\left\{u(z(t)) \exp\left[-\int_0^t \left(\frac{u''}{2u}\right)(z(s))\, ds\right]\right\} = u(x). \tag{2.8}$$

Since $u(y) \geq \alpha > 0$ for all y,

$$E_x\left\{\exp\left[-\int_0^t \left(\frac{u''}{2u}\right)(z(s))\, ds\right]\right\} \leq \frac{u(x)}{\alpha}.$$

But

$$E_x\left\{\exp\left[-\int_0^t \left(\frac{u''}{2u}\right)(z(s))\, ds\right]\right\} = E_x\left\{\exp\left[-t\int_{-\infty}^\infty \left(\frac{u''}{2u}\right)(y)\, dL(t, \omega, y)\right]\right\}$$

$$= E^{Q_{x,t}}\left\{\exp\left[-t\int_{-\infty}^\infty \left(\frac{u''}{2u}\right)(y)\, dF(y)\right]\right\},$$

and therefore for any measurable $C \subset \mathcal{F}$,

$$Q_{x,t}(C) \leq \frac{u(x)}{\alpha} \exp\left\{t \sup_{F \in C} \int_{-\infty}^\infty \left(\frac{u''}{2u}\right)(y)\, dF(y)\right\}. \tag{2.9}$$

Since (2.9) holds for any $u \in \mathcal{U}$, we obtain for any measurable $C \subset \mathcal{F}$,

$$\overline{\lim_{t \to \infty}} \frac{1}{t} \ln Q_{x,t}(C) \leq \inf_{u \in \mathcal{U}} \sup_{F \in C} \int_{-\infty}^\infty \left(\frac{u''}{2u}\right)(y)\, dF(y). \tag{2.10}$$

Let

$$I(F) = -\inf_{u \in \mathcal{U}} \int_{-\infty}^\infty \frac{u''(y)}{2u(y)}\, dF(y).$$

We see that $I(F)$ is lower semicontinuous on $\mathcal{F}$ and $0 \leq I(F) \leq \infty$. We will show now that for compact sets $C \subset \mathcal{F}$ it follows from (2.10) that

$$\overline{\lim_{t \to \infty}} \frac{1}{t} \ln Q_{x,t}(C) \leq \sup_{F \in C} \inf_{u \in \mathcal{U}} \int_{-\infty}^\infty \frac{u''(y)}{2u(y)}\, dF(y)$$

$$= -\inf_{f \in C} I(F)$$

First of all, for any measurable set $C \subset \mathcal{F}$ which is contained in a finite union $\bigcup_{j=1}^{k} C_j$ of measurable sets, we have from (2.10)

$$\overline{\lim_{t \to \infty}} \frac{1}{t} \ln Q_{x,t}(C) \leq \max_{1 \leq j \leq k} \inf_{u \in \mathcal{U}} \sup_{F \in C_j} \int_{-\infty}^{\infty} \frac{u''(y)}{2u(y)} dF(y),$$

and for any measurable set $C \subset \mathcal{F}$

$$\overline{\lim_{t \to \infty}} \frac{1}{t} \ln Q_{x,t}(C) \leq \inf_{\substack{C_1, C_2, \ldots, C_k \\ C \subset \bigcup_{j=1}^{k} C_j}} \max_{1 \leq j \leq k} \inf_{u \in \mathcal{U}} \sup_{F \in C_j} \int_{-\infty}^{\infty} \frac{u''(y)}{2u(y)} dF(y).$$

Let C be a compact subset of $\mathcal{F}$ and let

$$l = \sup_{F \in C} \inf_{u \in \mathcal{U}} \int_{-\infty}^{\infty} \frac{u''(y)}{2u(y)} dF(y),$$

so that for every $F \in C$,

$$\inf_{u \in \mathcal{U}} \int_{-\infty}^{\infty} \frac{u''(y)}{2u(y)} dF(y) \leq l.$$

Thus, given $\varepsilon > 0$, for every $F \in C$ there exists $u_F \in \mathcal{U}$ such that

$$\int_{-\infty}^{\infty} \frac{u_F''(y)}{2u_F(y)} dF(y) \leq l + \varepsilon.$$

Since $u''(y)/2u(y)$ is bounded and continuous on $-\infty < y < \infty$ for every $u \in \mathcal{U}$, $\int_{-\infty}^{\infty} u_F''(y)/2u_F(y) dF(y)$ is a continuous function on $\mathcal{F}$. Thus for every $F \in C$ there is a Levy neighbourhood N_F such that for every $G \in N_F$,

$$\int_{-\infty}^{\infty} \frac{u_F''(y)}{2u_F(y)} dG(y) \leq l + \varepsilon.$$

Now $\bigcup_{F \in C} N_F$ is an open covering of C and therefore there exists $F_1, F_2, \ldots, F_k$ in C such that $C \subset \bigcup_{j=1}^{k} N_{F_j}$. Let $C_j = N_{F_j}$ and we have

$$\sup_{G \in C_j} \int_{-\infty}^{\infty} \frac{u_{F_j}''(y)}{2u_{F_j}(y)} dG(y) \leq l + 2\varepsilon,$$

$$\inf_{u \in \mathcal{U}} \sup_{G \in C_j} \int_{-\infty}^{\infty} \frac{u''(y)}{2u(y)} dG(y) \leq l + 2\varepsilon,$$

$$\max_{1 \leq j \leq k} \inf_{u \in \mathcal{U}} \sup_{G \in C_j} \int_{-\infty}^{\infty} \frac{u''(y)}{2u(y)} dG(y) \leq l + 2\varepsilon,$$

and finally

$$\inf_{\substack{C_1,C_2,\ldots,C_k \\ \bigcup_{j=1}^{k} C_j \supset C}} \max_{1 \leq j \leq k} \inf_{u \in \mathcal{U}} \sup_{G \in C_j} \int_{-\infty}^{\infty} \frac{u''(y)}{2u(y)} dG(y) \leq l + 2\varepsilon.$$

Since ε is arbitrary, we conclude that for compact sets C†

$$\inf_{\substack{C_1,C_2,\ldots,C_k \\ \bigcup_{j=1}^{k} C_j \supset C}} \max_{1 \leq j \leq k} \inf_{u \in \mathcal{U}} \sup_{F \in C_j} \int_{-\infty}^{\infty} \frac{u''(y)}{2u(y)} dF(y) \leq \sup_{F \in C} \inf_{u \in \mathcal{U}} \int_{-\infty}^{\infty} \frac{u''(y)}{2u(y)} dF(y)$$

$$= -\inf_{F \in C} I(F). \qquad (2.13)$$

Thus (2.13) and (2.12) imply (2.11).

For $0 < K < \infty$, let $A_K = \{F \in \mathcal{F} : \Phi(F) \leq K\}$ which is compact from hypothesis (3) on Φ. From (2.1) we have

$$E_x\{e^{-t\Phi\{L(t,\omega,\cdot)\}}\} = E_{A_K}^{Q_{x,t}}\{e^{-t\Phi(F)}\} + E_{A_K^c}^{Q_{x,t}}\{e^{-t\Phi(F)}\}$$
$$\leq E_{A_K}^{Q_{x,t}}\{e^{-t\Phi(F)}\} + e^{-tK}. \qquad (2.14)$$

Let $F^* \in A_K$ and let $\varepsilon > 0$ be given. Assume first of all that $I(F^*) > \infty$. Then, since both $\Phi(F)$ and $I(F)$ are lower semicontinuous at F^*, there exists a neighbourhood N_{F^*} such that for all $F \in N_{F^*}$,

$$\Phi(F^*) \leq \Phi(F) + \varepsilon,$$
$$I(F^*) \leq I(F) + \varepsilon.$$

Let G_{F^*} be another neighbourhood of F^* such that $\bar{G}_{F^*} \subset N_{F^*}$. We have

$$E_{\bar{G}_{F^*} \cap A_K}^{Q_{x,t}}(e^{-t\Phi(F)}) \leq e^{-t[\Phi(F^*) - \varepsilon]} Q_{x,t}(\bar{G}_{F^*} \cap A_K).$$

Since $\bar{G}_{F^*} \cap A_K$ is a closed subset of a compact set it is itself compact, and therefore from (2.11)

$$\overline{\lim_{t \to \infty}} \frac{1}{t} \ln E_{\bar{G}_{F^*} \cap A_K}^{Q_{x,t}}(e^{-t\Phi(F)}) \leq -\Phi(F^*) + \varepsilon - \inf_{F \in \bar{G}_{F^*} \cap A_K} I(F)$$

$$\leq -\Phi(F^*) + \varepsilon - [I(F^*) - \varepsilon]$$

$$\leq -\inf_{F \in \mathcal{F}} [\Phi(F) + I(F)] + 2\varepsilon. \qquad (2.15)$$

On the other hand, if $I(F^*) = \infty$ we have from the lower semicontinuity of $I(F)$ that there exists a neighbourhood M_{F^*} such that for all $F \in M_{F^*}$, $I(F) \geq K$. Let G_{F^*} be another neighbourhood of F^* such that $\bar{G}_{F^*} \subset M_{F^*}$.

† In fact it is easy to show that for compact sets C equality holds in (2.13).

Then
$$E^{Q_{x,t}}_{\bar{G}_{F^*} \cap A_K}(e^{-t\Phi(F)}) \leq Q_{x,t}(\bar{G}_{F^*} \cap A_K),$$

and again from (2.11) we have

$$\overline{\lim_{t\to\infty}} \frac{1}{t} \ln E^{Q_{x,t}}_{\bar{G}_{F^*} \cap A_K}(e^{-t\Phi(F)}) \leq - \inf_{F \in \bar{G}_{F^*} \cap A_K} I(F) \leq -K. \tag{2.16}$$

Thus each point F^* of A_K is covered by a neighbourhood G_{F^*} and since A_K is compact there exists $F_1^*, F_2^*, \ldots, F_n^*$ such that $A_K \subset \bigcup_{j=1}^n G_{F_j^*}$. From (2.15) and (2.16) we have, therefore,

$$\overline{\lim_{t\to\infty}} \frac{1}{t} \ln E^{Q_{x,t}}_{A_K}(e^{-t\Phi(F)}) \leq -\min\{K, \inf_{F \in \mathcal{F}} [\Phi(F) + I(F)] - 2\varepsilon\}. \tag{2.17}$$

From (2.14) and (2.17) we obtain

$$\overline{\lim_{t\to\infty}} \frac{1}{t} \ln E_x(e^{-t\Phi[L(t,\omega,\cdot)]}) \leq -\min\{K, \inf_{F \in \mathcal{F}} [\Phi(F) + I(F)] - 2\varepsilon\},$$

and letting $\varepsilon \to 0$ we get

$$\overline{\lim_{t\to\infty}} \frac{1}{t} \ln E_x(e^{-t\Phi[L(t,\omega,\cdot)]}) \leq -\min\{K, \inf_{F \in \mathcal{F}} [\Phi(F) + I(F)]\}.$$

Finally, letting $K \to \infty$ we find

$$\overline{\lim_{t\to\infty}} \frac{1}{t} \ln E_x(e^{-t\Phi[L(t,\omega,\cdot)]}) \leq - \inf_{F \in \mathcal{F}} [\Phi(F) + I(F)]. \tag{2.18}$$

In the next section we obtain a lower estimate on

$$\lim_{t\to\infty} \frac{1}{t} \ln E_x(e^{-t\Phi[L(t,\omega,\cdot)]}),$$

and in § 2.4 we show that this lower estimate and the upper estimate on the right of (2.18) are equal and in fact are the same as the right side of (2.4). This will prove the theorem.

2.3. The lower estimate

Let $[a, b]$ be a finite closed interval such that $x \in (a, b)$. Let $f(y)$ be a density function having two continuous derivatives on $-\infty < y < \infty$, $f(y) > 0$ for $y \in (a, b)$, $f(y) = 0$ for y outside of $[a, b]$, and

$$\int_a^b \frac{[f'(y)]^2}{f(y)} dy < \infty.$$

Let $F(y)$ be the distribution function corresponding to $f(y)$.

For each $\omega = z(\,.\,) \in \Omega_x$ such that $a \leq z(s) \leq b$ for all $0 \leq s \leq t$, $L(t, \omega, y)$ is a distribution function whose support is contained in $[a, b]$. Thus, by hypothesis (4) (p.16) on Φ, for any $\varepsilon > 0$ there exists a Lévy neighbourhood of F, say N_F, so that if $L(t, \omega, \,.\,) \in N_F$ for any ω such that $a \leq z(s) \leq b$ for all $0 \leq s \leq t$, then

$$|\Phi(L(t, \omega, \,.\,)) - \Phi(F)| \leq \varepsilon.$$

Thus, if we let $C^t_{a,b} = \{z \in \Omega_x : a \leq z(s) \leq b \text{ for all } 0 \leq s \leq t\}$,

$$E_x(e^{-t\Phi\{L(t,\omega,\cdot)\}}) \geq E_x\{e^{-t\Phi[L(t,\omega,\cdot)]}; L(t, \omega, \,.\,) \in N_F, z \in C^t_{a,b}\}$$
$$\geq e^{-t[\Phi(F)+\varepsilon]} P_x\{L(t, \omega, \,.\,) \in N_F, z \in C^t_{a,b}\}. \quad (2.19)$$

In Lemma 2.1, which follows, we will show

$$\lim_{t \to \infty} \frac{1}{t} \ln P_x\{L(t, \omega, \,.\,) \in N_F, z \in C^t_{a,b}\} \geq -\frac{1}{8} \int_a^b \frac{[f'(y)]^2}{f(y)} \, dy. \quad (2.20)$$

From (2.19) we obtain that

$$\lim_{t \to \infty} \frac{1}{t} \ln E_x\{e^{-t\Phi\{L(t,\omega,\cdot)\}}\} \geq -\Phi(F) - \varepsilon - \frac{1}{8} \int_a^b \frac{[f'(y)]^2}{f(y)} \, dy,$$

and since ε is arbitrary we obtain

$$\lim_{t \to \infty} \frac{1}{t} \ln E_x(e^{-t\Phi\{L(t,\omega,\cdot)\}}) \geq -\left\{\Phi(F) + \frac{1}{8} \int_a^b \frac{[f'(y)]^2}{f(y)} \, dy\right\}$$
$$\geq -\inf_{f \in f_x} \left\{\Phi(f) + \frac{1}{8} \int_a^b \frac{[f'(y)]^2}{f(y)} \, dy\right\}, \quad (2.21)$$

where f_x is the set of all density functions f with two continuous derivatives for which there is a closed interval $[a, b]$ containing x in its interior, outside of which $f = 0$, $f > 0$ on (a, b), and

$$\int_a^b \frac{[f'(y)]^2}{f(y)} \, dy < \infty.$$

Since (2.20) is of independent interest we prove it as a lemma.

LEMMA 2.1. *Let $f \in f_x$ and let N_F be any Lévy neighbourhood of its corresponding distribution function. Then*

$$\lim_{t \to \infty} \frac{1}{t} \ln P_x\{L(t, \omega, \,.\,) \in N_F, a \leq z(s) \leq b \text{ for all } 0 \leq s \leq t\}$$
$$\geq -\frac{1}{8} \int_a^b \frac{[f'(y)]^2}{f(y)} \, dy.$$

Proof. Let $b(y) = \frac{1}{2} f'(y)/f(y)$ and consider Brownian motion with drift b, i.e. the process whose generator is

$$\frac{1}{2} \frac{\partial^2}{\partial x^2} + b(x) \frac{\partial}{\partial x}.$$

Now, the solution of

$$\frac{1}{2} \frac{\partial^2 u}{\partial x^2} + b(x) \frac{\partial u}{\partial x} = 0$$

being, in the case $b(x) = \frac{1}{2} f'(x)/f(x)$, simply $u' = 1/f$, we have

$$u(y) = \int_\eta^y \frac{d\xi}{f(\xi)}$$

for $y \in (a, b)$, $\eta \in (a, b)$. Hence $u(y) \to -\infty$ as $y \to a$ and $u(y) \to +\infty$ as $y \to b$, which means according to Feller's test [2] that with probability 1 Brownian-motion paths with this drift never leave (a, b).

If we let $\tilde{P}_x$ denote the measure on Ω_x induced by this drift process, we have the Cameron–Martin formula [3]

$$P_x\{A \cap [z(s) \in [a, b] \text{ for all } 0 \le s \le t]\}$$
$$= E_A^{\tilde{P}_x}\left\{\exp\left[-\int_0^t b(z(s))\, dz(s) + \frac{1}{2} \int_0^t b^2(z(s))\, ds\right]\right\}.$$

In our case, this becomes

$$P_x\{L(t, \omega, .) \in N_F,\ z(s) \in [a, b] \text{ for all } 0 \le s \le t\}$$
$$= E^{\tilde{P}_x}\left\{\exp\left[-\frac{1}{2} \int_0^t \left(\frac{f'}{f}\right)(z(s))\, dz(s) + \frac{1}{8} \int_0^t \left(\frac{f'}{f}\right)^2 (z(s))\, ds\right];\ L(t, \omega, .) \in N_F\right\}. \qquad (2.22)$$

Since f is twice continuously differentiable, we have by Ito's formula

$$d \ln f(z(s)) = (\ln f)'(z(s))\, dz(s) + \frac{1}{2} (\ln f)''(z(s))\, ds,$$

and, integrating this from 0 to t using $z(0) = x$, we obtain

$$\ln f(z(t)) - \ln f(x)$$
$$= \int_0^t \left(\frac{f'}{f}\right)(z(s))\, dz(s) + \frac{1}{2} \int_0^t \left(\frac{f''}{f}\right)(z(s))\, ds - \frac{1}{2} \int_0^t \left(\frac{f'}{f}\right)^2 (z(s))\, ds. \qquad (2.23)$$

Using (2.23) in (2.22) we obtain

$$P_x\{L(t, \omega, \cdot) \in N_F, z(s) \in [a, b] \text{ for all } 0 \le s \le t\}$$
$$= E^{\tilde{P}_x}\left(\sqrt{\left[\frac{f(x)}{f(z(t))}\right]}\exp\left\{-\left[\frac{1}{8}\int_0^t \left(\frac{f'}{f}\right)^2(z(s))\,ds - \frac{1}{4}\int_0^t \left(\frac{f''}{f}\right)(z(s))\,ds\right]\right\};$$
$$L(t, \omega, \cdot) \in N_F\right). \tag{2.24}$$

Let

$$J(y) = \frac{1}{8}\left(\frac{f'}{f}\right)^2(y) - \frac{1}{4}\left(\frac{f''}{f}\right)(y),$$

and let $\varepsilon > 0$ be given. Define

$$S(t, \varepsilon) = \left\{z \in \Omega_x : \left|\int_a^b J(y)\,dL(t, \omega, y) - \int_a^b J(y)f(y)\,dy\right| < \varepsilon\right\},$$

and

$$S'(t, \varepsilon) = \{z \in \Omega_x : L(t, \omega, \cdot) \in N_F\} \cap S(t, \varepsilon).$$

Since, for almost all $z \in \Omega_x$ ($\tilde{P}_x$-measure),

$$\int_a^b J(y)\,dL(t, \omega, y) = \int_{-\infty}^\infty J(y)L(t, \omega, y) = \frac{1}{t}\int_0^t J(z(s))\,ds,$$

we have from (2.24)

$$P_x\{L(t, \omega, \cdot) \in N_F, z(s) \in [a, b] \text{ for all } 0 \le s \le t\}$$
$$= E^{\tilde{P}_x}\left(\sqrt{\left[\frac{f(x)}{f(z(t))}\right]}\exp\left\{-t\int_a^b J(y)L(t, \omega, y)\right\}; L(t, \omega, \cdot) \in N_F\right)$$
$$\ge \exp\left\{-t\left[\varepsilon + \int_a^b J(y)f(y)\,dy\right]\right\}E^{\tilde{P}_x}\left\{\sqrt{\left[\frac{f(x)}{f(z(t))}\right]}; S'(t, \varepsilon)\right\}$$
$$= \exp\left\{-\varepsilon t - \frac{t}{8}\int_a^b \frac{[f'(y)]^2}{f(y)}\,dy\right\}E^{\tilde{P}_x}\left\{\sqrt{\left[\frac{f(x)}{f(z(t))}\right]}; S'(t, \varepsilon)\right\}. \tag{2.25}$$

Because of the hypotheses on f there exists a constant γ such that

$$\sqrt{\left[\frac{f(x)}{f(z(t))}\right]} \ge \gamma > 0$$

for almost all $z \in \Omega_x$ ($\tilde{P}_x$-measure). Therefore

$$E^{\tilde{P}_x}\left\{\sqrt{\left[\frac{f(x)}{f(z(t))}\right]};\ S'(t,\varepsilon)\right\} \geq \gamma \tilde{P}_x(S'(t,\varepsilon))$$

$$= \gamma - \gamma \tilde{P}_x(\Omega_x - S'(t,\varepsilon)). \quad (2.26)$$

Now the reason the drift $b(y) = \frac{1}{2}f'(y)/f(y)$ was selected is because then f itself is the invariant measure for the drift process. This is easy to see since if the drift is $\frac{1}{2}f'/f$ then $u = f$ is the solution of

$$\tfrac{1}{2}u'' - [b(x)u(x)]' = 0,$$

the adjoint equation. From the ergodic properties of Markov processes (see e.g. [2])

$$\lim_{t \to \infty} \tilde{P}_x(\Omega_x - S(t,\varepsilon))$$

$$= \lim_{t \to \infty} \tilde{P}_x\left(z \in \Omega_x : \left|\int_a^b J(y)\,dL(t,\omega,y) - \int_a^b J(y)f(y)\,dy\right| \geq \varepsilon\right)$$

$$= 0$$

and

$$\lim_{t \to \infty} \tilde{P}_x(z \in \Omega_x : L(t,\omega,\,.\,) \notin N_F) = 0.$$

These last two statements imply

$$\lim_{t \to \infty} \tilde{P}_x(\Omega_x - S'(t,\varepsilon)) = 0,$$

and, therefore, from (2.25) and (2.26) we obtain

$$\lim_{t \to \infty} \frac{1}{t} \ln P_x\{L(t,\omega,\,.\,) \in N_F,\ z(s) \in [a,b]\ \text{ for all }\ 0 \leq s \leq t\}$$

$$\geq -\varepsilon - \frac{1}{8}\int_a^b \frac{[f'(y)]^2}{f(y)}\,dy.$$

Finally, letting $\varepsilon \to 0$, we obtain (2.20).

2.4. The equality of the estimates

In this section we want to show that the expressions on the right of (2.18) and (2.21) are equal and the same as the expression on the right of (2.4). We shall first show Lemma 2.2.

LEMMA 2.2. *Let f be a density function such that $f > 0$ on (a, b) where $-\infty \leq a < b \leq +\infty$, $f = 0$ outside of $[a, b]$ and f is continuously differentiable on $-\infty < y < \infty$. Then, letting F be the corresponding distribution function,*

$$I(F) = \frac{1}{8} \int_a^b \frac{[f'(y)]^2}{f(y)} \, dy, \tag{2.27}$$

in the sense that if one side is finite so is the other and they are equal.

Proof. Assume first that

$$\int_a^b \frac{[f'(y)]^2}{f(y)} \, dy < \infty.$$

We show now that for every $u \in \mathcal{U}$

$$\int_{-\infty}^{\infty} \frac{u''(y)}{2u(y)} \, dF(y) \geq -\frac{1}{8} \int_a^b \frac{[f'(y)]^2}{f(y)} \, dy, \tag{2.28}$$

which will imply

$$I(F) \leq \frac{1}{8} \int_a^b \frac{[f'(y)]^2}{f(y)} \, dy < \infty. \tag{2.29}$$

So, for any $u \in \mathcal{U}$, let $h = \ln u$ and we have

$$\int_{-\infty}^{\infty} \frac{u''(y)}{2u(y)} \, dF(y) = \frac{1}{2} \int_{-\infty}^{\infty} \{h''(y) + [h'(y)]^2\} f(y) \, dy$$

$$= \frac{1}{2} \int_a^b \{h''(y) + [h'(y)]^2\} f(y) \, dy. \tag{2.30}$$

Using elementary inequalities

$$\left| \int_a^b h'(y) f'(y) \, dy \right| = \left| \int_a^b \frac{h'(y) f'(y)}{\sqrt{[f(y)]}} \sqrt{[f(y)]} \, dy \right|$$

$$\leq \sqrt{\left[\int_a^b [h'(y)]^2 f(y) \, dy \right]} \sqrt{\left[\int_a^b \frac{[f'(y)]^2}{f(y)} \, dy \right]}$$

$$= 2 \sqrt{\left[\int_a^b [h'(y)]^2 f(y) \, dy \right]} \sqrt{\left[\frac{1}{4} \int_a^b \frac{[f'(y)]^2}{f(y)} \, dy \right]}$$

$$\leq \int_a^b [h'(y)]^2 f(y) \, dy + \frac{1}{4} \int_a^b \frac{[f'(y)]^2}{f(y)} \, dy,$$

and hence

$$\frac{1}{2}\left\{\int_a^b [h'(y)]^2 f(y)\,dy - \int_a^b h'(y)f'(y)\,dy\right\} \geq -\frac{1}{8}\int_a^b \frac{[f'(y)]^2}{f(y)}\,dy.$$

Integrating by parts we obtain (2.28) and hence (2.29).

To complete the proof it suffices to show that if F is such that $I(F)<\infty$, then

$$\frac{1}{8}\int_a^b \frac{[f'(y)]^2}{f(y)}\,dy \leq I(F). \tag{2.31}$$

From the definition of $I(F)$, we have for every $u \in \mathcal{U}$ and every $h = \ln u$

$$\int_a^b \frac{u''(y)}{2u(y)}\,dF(y) = \frac{1}{2}\int_a^b [h'(y)]^2 f(y)\,dy + \frac{1}{2}\int_a^b h''(y)f(y)\,dy \geq -I(F). \tag{2.32}$$

Now (2.32) holds in particular for these $u \in \mathcal{U}$ such that $h = \ln u$ are constant outside of a closed bounded interval contained in (a, b). If we let $\mathcal{H}$ denote this latter class of functions h, we have from (2.32) for any $h \in \mathcal{H}$

$$\frac{1}{2}\int_a^b [h'(y)]^2 f(y)\,dy - \frac{1}{2}\int_a^b h'(y)f'(y)\,dy \geq -I(F). \tag{2.33}$$

If $h \in \mathcal{H}$ then $\lambda h \in \mathcal{H}$ for any real λ, so from (2.33) we have for all $h \in \mathcal{H}$

$$\frac{\lambda^2}{2}\int_a^b [h'(y)]^2 f(y)\,dy - \frac{\lambda}{2}\int_a^b h'(y)f'(y)\,dy + I(F) \geq 0.$$

Thus, the discriminant of this quadratic in λ must be less than or equal to zero, i.e. for all $h \in \mathcal{H}$,

$$\left|\int_a^b h'(y)f'(y)\,dy\right| \leq \sqrt{(8I(F))}\sqrt{\left\{\int_a^b [h'(y)]^2 f(y)\,dy\right\}}. \tag{2.34}$$

If we let $g(y) = h'(y)$, then we have

$$\left|\int_a^b g(y)f'(y)\,dy\right| \leq \sqrt{[8I(F)]}\sqrt{\left[\int_a^b g^2(y)f(y)\,dy\right]} \tag{2.35}$$

is valid for all functions $g(y)$ which have compact support contained in (a, b) and which have a continuous derivative. By approximation we then have (2.35) valid for all functions $g(y)$ which are continuous and have compact support contained in (a, b). Let $\phi(y) \in C^\infty$ with compact support contained in (a, b) and $0 \leq \phi(y) \leq 1$ for all $-\infty < y < \infty$. We now apply (2.35) to

$g(y) = [f'(y)/f(y)]\phi(y)$, obtaining

$$\left| \int_a^b \frac{[f'(y)]^2}{f(y)} \phi(y)\, dy \right| \leq \sqrt{(8I(F))} \sqrt{\left\{ \int_a^b \frac{[f'(y)]^2}{f(y)} \phi^2(y)\, dy \right\}}$$

$$\leq \sqrt{(8I(F))} \sqrt{\left\{ \int_a^b \frac{[f'(y)]^2}{f(y)} \phi(y)\, dy \right\}}.$$

Dividing and squaring we obtain

$$\frac{1}{8} \int_a^b \frac{[f'(y)]^2}{f(y)} \phi(y)\, dy \leq I(F). \tag{2.36}$$

Since (2.36) holds for all $\phi(y)$ such that $0 \leq \phi(y) \leq 1$, $\phi(y) \in C^\infty$, $\phi(y)$ has compact support contained in (a, b) we conclude that in fact

$$\frac{1}{8} \int_a^b \frac{[f'(y)]^2}{f(y)}\, dy \leq I(F).$$

We now prove Lemma 2.3.

LEMMA 2.3.

$$\inf_{f \in f_x} \left\{ \frac{1}{8} \int_a^b \frac{[f'(y)]^2}{f(y)} dy + \Phi(f) \right\} = \inf_{F \in \mathcal{F}} [I(F) + \Phi(F)]. \tag{2.37}$$

Proof. Let $f \in f_x$, then from Lemma 2.2 we have

$$\frac{1}{8} \int_a^b \frac{[f'(y)]^2}{f(y)} dy = I(F)$$

where F is the distribution function with f. Thus, the left side of (2.37) is clearly greater than or equal to the right side. We prove the reverse inequality in successive stages.

Let $\mathcal{F}_c$ be the class of distribution functions having compact support. We show now

$$\inf_{F \in \mathcal{F}_c} [I(F) + \Phi(F)] = \inf_{F \in \mathcal{F}} [I(F) + \Phi(F)]. \tag{2.38}$$

It suffices to show the left side of (2.38) is less than or equal to the right side, and for this let $F \in \mathcal{F}$ such that $\Phi(F) = a_1 < \infty$ and $I(F) = a_2 < \infty$. We then show that for any $\varepsilon > 0$ there exists a distribution function $G \in \mathcal{F}_c$ such that $\Phi(G) < a_1 + \varepsilon$ and $I(G) < a_2 + \varepsilon$.

Define a sequence of functions $\{h_n(y)\}$ such that $0 \leq h_n(y) \leq 1$ for $-\infty < y < \infty$, $h_n(h) = 1$ if $|y| \leq n$, $h_n(y) = 0$ if $|y| > 2n$, and $|h'_n(y)| < 1/n$ for $n < |y| \leq 3n$. For the given $F \in \mathcal{F}$ and $g_n(y) = h_n^2(y)$, define

$$F_n(y) = \frac{\int_{-\infty}^{y} g_n(\xi) \, dF(\xi)}{\int_{-\infty}^{\infty} g_n(\xi) \, dF(\xi)}.$$

For each n, $F_n \in \mathcal{F}_c$ and from hypothesis (5) (p. 16) on $\Phi(F)$, there exists an n_0 so that $n \geq n_0$ implies $\Phi(F_n) \leq \Phi(F) + \varepsilon = a_1 + \varepsilon$.

We show now that $\lim_{n \to \infty} I(F_n) = I(F)$. First, since I is lower semicontinuous on $\mathcal{F}$, $\underline{\lim}_{n \to \infty} I(F_n) \geq I(F)$, and so we want to show $\overline{\lim}_{n \to \infty} I(F_n) \leq I(F)$. For any $\eta > 0$, let ψ_η be the Gaussian distribution with mean 0 and variance η. For our given $F \in \mathcal{F}$, define $F_\eta = F * \psi_\eta$. By convexity†, $I(F_\eta) \leq I(F)$, and moreover, since F_η has a strictly positive density f_η with continuous derivatives, we have from Lemma 2.2,

$$I(F_\eta) = \frac{1}{8} \int_{-\infty}^{\infty} \frac{[f'_\eta(y)]^2}{f_\eta(y)} dy. \tag{2.39}$$

Let

$$F_{n,\eta}(y) = \frac{\int_{-\infty}^{y} g_n(\xi) \, dF_\eta(\xi)}{\int_{-\infty}^{\infty} g_n(\xi) \, dF_\eta(\xi)},$$

and note $F_{n,\eta}(y)$ has a strictly positive density with continuous derivative, namely,

$$\frac{g_n f_\eta}{\int_{-\infty}^{\infty} g_n(\xi) f_\eta(\xi) \, d\xi}.$$

Thus again from Lemma 2.2, and the definition of g_n,

$$I(F_{n,\eta}) = \frac{1}{8 \int_{-\infty}^{\infty} g_n(\xi) f_\eta(\xi) \, d\xi} \int_{-\infty}^{\infty} \frac{[f_\eta(y) g_n(y)']^2}{f_\eta(y) g_n(y)} dy$$

$$= \frac{1}{8 \int_{-\infty}^{\infty} g_n(\xi) f_\eta(\xi) \, d\xi} \left\{ \int_{-\infty}^{\infty} \frac{[f'_\eta(y)]^2}{f_\eta(y)} g_n(y) \, dy + \right.$$

$$\left. + \int_{-\infty}^{\infty} f_\eta(y) \frac{[g'_n(y)]^2}{g_n(y)} dy + 2 \int_{-\infty}^{\infty} f'_\eta(y) g'_n(y) \, dy \right\}$$

$$\leq \frac{1}{8 \int_{-\infty}^{\infty} g_n(\xi) f_\eta(\xi) d\xi} \left\{ \int_{-\infty}^{\infty} \frac{[f'_\eta(y)]^2}{f_\eta(y)} dy + \frac{4}{n^2} \int_{-\infty}^{\infty} f_\eta(y) \, dy \right.$$

$$\left. + \frac{2}{n} \int_{-\infty}^{\infty} |f'_\eta(y)| dy \right\}.$$

† From the definition of $I(F)$ it follows that it is translation invariant, is lower semicontinuous, and is convex. Therefore it decreases under convolution.

From (2.39) and the fact, noted earlier, that $I(F_n) \leq I(F)$,

$$\int_{-\infty}^{\infty} |f'_n(y)| \, dy = \int_{-\infty}^{\infty} \frac{|f'_n(y)|}{\sqrt{[f_n(y)]}} \sqrt{[f_n(y)]} \, dy$$

$$\leq \sqrt{\left\{\int_{-\infty}^{\infty} \frac{[f'_n(y)]^2}{f_n(y)} \, dy\right\}} \sqrt{\left\{\int_{-\infty}^{\infty} f_n(y) \, dy\right\}}$$

$$= \sqrt{\left\{\int_{-\infty}^{\infty} \frac{[f'_n(y)]^2}{f_n(y)}\right\} \, dy} = \sqrt{[8I(F_n)]} \leq \sqrt{[8I(F)]}.$$

Therefore, again using (2.39), and $I(F_n) \leq I(F)$,

$$I(F_{n,\eta}) \leq \frac{1}{8 \int_{-\infty}^{\infty} g_n(\xi) f_\eta(\xi) \, d\xi} \left\{ 8I(F) + \frac{4}{n^2} + \frac{2}{n} \sqrt{[8I(F)]} \right\}. \quad (2.40)$$

Holding n fixed and letting $\eta \to 0$ we see that

$$\lim_{\eta \to 0} I(F_{n,\eta}) = I(F_n),$$

$$\lim_{\eta \to 0} \int_{-\infty}^{\infty} g_n(\xi) f_\eta(\xi) \, d\xi = \int_{-\infty}^{\infty} g_n(\xi) \, d\xi,$$

and therefore from (2.40), letting $\eta \to 0$,

$$I(F_n) \leq \frac{1}{8 \int_{-\infty}^{\infty} g_n(\xi) \, d\xi} \left\{ 8I(F) + \frac{4}{n^2} + \frac{2}{n} \sqrt{[8I(F)]} \right\}.$$

Thus $\overline{\lim}_{n \to \infty} I(F_n) \leq I(F)$, and so there exists an n_1 such that $n > n_1$ implies $I(F_n) \leq I(F) + \varepsilon = a_2 + \varepsilon$. For any $n > \max(n_0, n_1)$ choose $G(y) = F_n(y) \in \mathcal{F}_c$ and $\Phi(G) < a_1 + \varepsilon$ and $I(G) < a_2 + \varepsilon$ so that we have (2.38).

Now let $\mathcal{F}_{c'}$ be the space of distribution functions having compact support and possessing twice continuously differentiable density functions. We show now

$$\inf_{F \in \mathcal{F}_{c'}} [I(F) + \Phi(F)] = \inf_{F \in \mathcal{F}_c} [I(F) + \Phi(F)]. \quad (2.41)$$

Again, of course, it suffices to show the left side of (2.41) is less than or equal to the right side. Let $F \in \mathcal{F}_c$. For any $\varepsilon > 0$ we want to find $G \in \mathcal{F}_{c'}$ such that $I(G) < I(F) + \varepsilon$ and $\Phi(G) < \Phi(F) + \varepsilon$. For any $\eta > 0$, let $\eta(y)$ be a density function which is twice continuously differentiable with compact support $\subset (-\eta, \eta)$. For our $F \in \mathcal{F}_c$, define

$$g_\eta(x) = \int \eta(x - y) \, dF(y).$$

Thus $g_\eta(x)$ is a density function which is twice continuously differentiable and has compact support, since both $\eta(y)$ and $F(y)$ have compact support. If we let G_η be the distribution function corresponding to the density g_η we have that for all $\eta > 0$, $G_\eta \in \mathcal{F}_c$. Moreover, by convexity, for all $\eta > 0$,

$$I(G_\eta) \leq I(F) \leq I(F) + \varepsilon.$$

On the other hand, as $\eta \to 0$, $G_\eta \Rightarrow F$ and the support of G_η is for every $\eta > 0$ contained in a fixed interval independent of η. Thus by hypothesis (4) (p. 16) on Φ we can choose η so small that $\Phi(G_\eta) \leq \Phi(F) + \varepsilon$. Hence, for any η so small, let $G = G_\eta$ and we have $I(G) \leq I(F) + \varepsilon$ and $\Phi(G) \leq \Phi(F) + \varepsilon$, establishing (2.41).

Finally, we show

$$\inf_{f \in f_x} \left\{ \frac{1}{8} \int_a^b \frac{[f'(y)]^2}{f(y)} dy + \Phi(f) \right\} = \inf_{F \in \mathcal{F}_c} [I(F) + \Phi(F)]. \qquad (2.42)$$

Since $f_x \subset \mathcal{F}_c$ it again suffices to show that for $F^* \in \mathcal{F}_c$ and given $\varepsilon > 0$ there exists a density $f \in f_x$ such that

$$\frac{1}{8} \int_a^b \frac{[f'(y)]^2}{f(y)} dy \leq I(F^*) + \varepsilon$$

and

$$\Phi(f) \leq \Phi(F^*) + \varepsilon.$$

Let $h \in f_x$ such that the interval $[a, b] \ni x$ and contains the support of F^* on the interior of which $h > 0$. Let for any $\eta > 0$,

$$g_\eta = (1 - \eta)f^* + \eta h$$

where f^* is the density of F^*. Thus g_η has two continuous derivatives and $g_\eta > 0$ on (a, b). Again by convexity, for any $\eta > 0$

$$I(g_\eta) \leq (1 - \eta)I(F^*) + \eta I(h)$$

so by letting $\eta \to 0$ and using hypothesis (4) (p. 16) on Φ we have for small enough η,

$$I(g_\eta) \leq I(F^*) + \varepsilon$$

and

$$\Phi(g_\eta) \leq \Phi(F^*) + \varepsilon.$$

Also from Lemma 2.2, since $I(g_\eta) < \infty$, we have

$$\int_a^b \frac{[g'_\eta(y)]^2}{g_\eta(y)} dy < \infty.$$

This gives (2.42).

Now, since $f_x \subset f \subset \mathcal{F}$, we have finally that

$$\inf_{f \in f_x} \left\{ \frac{1}{8} \int_a^b \frac{[f'(y)]^2}{f(y)} dy + \Phi(f) \right\} = \inf_{f \in f} \left\{ \frac{1}{8} \int_{-\infty}^{\infty} \frac{[f'(y)]^2}{f(y)} dy + \Phi(f) \right\}$$

$$= \inf_{F \in \mathcal{F}} [I(F) + \Phi(F)].$$

This completes the proof of the theorem.

Acknowledgements

This research was supported by the Office of Naval Research, Contract N00014-67-A-0467-0031 and the National Science Foundation, Grant NSF-GP-41985.

References

1. KAC, M. On some connections between probability theory and differential and integral equations, *Proceedings of the 2nd Berkeley Symposium*, pp. 189–215 (1950).
2. ITO, K. and MCKEAN JR., H. P. *Diffusion processes and their sample paths*. Springer–Verlag, New York (1965).
3. CAMERON, R. H. and MARTIN, W. T. The transformation of Wiener integrals by non-linear transformations. *Trans. Am. math. Soc.* **75**, 552–75 (1953).

3. Covariant Schrödinger equations

J. S. DOWKER

3.1. Deriving the equations

THE question of what is the Schrödinger equation for a particle restricted to lie on a manifold M has been the subject of discussion since the earliest days of quantum mechanics ([61], [52], [62], [69], [9]; see also [30]). The general concensus is that, in the absence of potentials, the wave equation takes the intrinsic, covariant form (also see [10]):

$$i\frac{\partial \psi(q, t)}{\partial t} = -\frac{1}{2}\Delta_2 \psi(q, t) \tag{3.1}$$

where Δ_2 is the Laplace–Beltrami operator

$$\Delta_2 \equiv g^{-\frac{1}{2}}\partial_\alpha(g^{\alpha\beta}g^{\frac{1}{2}}\partial_\beta), \qquad g = \det|g_{\alpha\beta}|$$

and ψ is a scalar of weight zero. In terms of the covariant derivative, denoted by a double stroke $\|$, $\Delta_2 \psi = g^{\alpha\beta} \psi_{\|\alpha\|\beta}$ ($g_{\alpha\beta}$ is, of course, the metric on M and α, $\beta \ldots = 1, 2, \ldots r$). Δ_2 is a formally self-adjoint operator (e.g. [4], [27; p. 387], [7]).

Equation (3.1) can be written down without reference to a canonical quantization procedure but this question cannot be avoided in any complete treatment of quantum mechanics. The classical Lagrangian and Hamiltonian are

$$L = \tfrac{1}{2}g_{\alpha\beta}\dot{q}^\alpha\dot{q}^\beta + A_\alpha\dot{q}^\alpha; \qquad H = \tfrac{1}{2}g^{\alpha\beta}(p_\alpha - A_\alpha)(p_\beta - A_\beta) \tag{3.2}$$

with $p_\alpha = g_{\alpha\beta}\dot{q}^\beta + A_\alpha$, where for later generality we have added a vector potential term. Quantum mechanically, p and q become operators, $\hat{p}$ and $\hat{q}$, with commutation rules $[\hat{p}_\alpha, \hat{p}_\beta] = 0 = [\hat{q}^\alpha, \hat{q}^\beta]$, $[\hat{p}_\alpha, \hat{q}^\beta] = -i\delta_\alpha^\beta$. There are well-known problems when the manifold M is closed, compact. If we wish to ignore these we can take each q^α to have the range $-\infty$ to $+\infty$ and assume our manifold to be simple convex. (Although these restrictions are probably not essential, this point needs further investigation.)

The Schrödinger representation of the canonical commutation relation (CCR) is ([48] [10]):

$$\hat{q}^\alpha = q^\alpha, \qquad \hat{p}_\alpha = -i(\partial_\alpha + \tfrac{1}{4}\partial_\alpha \ln g).$$

(We note that $\hat{p}_\alpha$ is not represented by the covariant derivative, as is sometimes supposed. This replacement would yield $-\tfrac{1}{2}\Delta_2$ from H by direct substitution but would also give a non-zero value for $[\hat{p}_\alpha, \hat{p}_\beta]\hat{p}_\gamma\psi$ in disagreement with the CCR.)

In terms of $\hat{p}$ and $\hat{q}$ the quantum Hamiltonian corresponding to (3.1) is written

$$\hat{H}(\hat{p},\hat{q}) = \tfrac{1}{2} g^{-\frac{1}{4}} \hat{p}_\alpha g^{\alpha\beta} g^{\frac{1}{2}} \hat{p}_\beta g^{-\frac{1}{4}} = \tfrac{1}{2} \hat{p}_\alpha g^{\alpha\beta} \hat{p}_\beta + Q,$$

with

$$Q = -\tfrac{1}{2} g^{\frac{1}{4}} \Delta_2 g^{-\frac{1}{4}},$$

and amounts to a particular choice of ordering ps and qs, which is fixed essentially by the postulated covariance of the quantum theory.

At first sight it might be imagined that Feynman path-integral quantization would yield a unique ordering but closer investigation shows that this is unfounded (e.g. [8], [64], [40], [41], [31]). Of course if the Feynman quantization is arranged to be covariant then there is little else the Schrödinger equation can be but (3.1), apart, possibly, from the appearance of terms involving the curvature $R_{\alpha\beta\gamma\delta}$. Nevertheless, in view of the facts that Feynman quantization has been elevated to the status of a postulate and that path-integrals are used extensively in the quantization of non-linear gauge theories such as gravitation and chiral dynamics it is interesting to discuss Lagrangians of the type (3.2), from this angle.

In his basic paper on quantization on curved spaces, DeWitt[11] extended the work of Pauli[49] and Morette[45] and derived a path-integral formula for the propagator $\langle q'', t''|q', t'\rangle$ of eqn (3.1) as a folding together of the infinitesimal ('short-time') propagators, in the manner of Dirac and Feynman[20]. (A useful and attractive introduction to this and other questions will be found in reference [14].) The short-time propagator is, basically, the WKB expression of Van Vleck[66] consisting of an amplitude and a phase, $D^{\frac{1}{2}} e^{iS}$, where S is the classical action and D the Van Vleck determinant. In the limit when the mesh defined by the integration over the intermediate variables becomes sufficiently fine the $D^{\frac{1}{2}}$ factor is equivalent to a term in the exponent and the DeWitt expression is

$$\langle q'', t''|q', t'\rangle = \lim_{\substack{n\to\infty \\ \varepsilon\to 0}} (2\pi i\varepsilon)^{-\frac{1}{2}} \int \prod_1^{n-1} \left[\frac{g^{\frac{1}{2}}(q_j)\, dq_j}{(2\pi i\varepsilon)^{n/2}}\right] \exp\left\{i\sum_1^n [S(q_j t_j|q_{j-1} t_{j-1}) + \tfrac{1}{6} R(q_j)\varepsilon]\right\} \quad (\varepsilon = t_j - t_{j-1}); \tag{3.3}$$

S being the *classical* action,

$$S(q_j t_j|q_{j-1} t_{j-1}) = \int_{j-1}^{j} L(q,\dot{q})\, dt.$$

Equation (3.3) can be written *formally* as a sum-over-paths,

$$\langle q'', t''|q', t'\rangle = \mathcal{N} \int_{q'}^{q''} \exp\left[i \int_{t'}^{t''} (L + \tfrac{1}{6} R)\, dt\right] \mathcal{D}[q]. \tag{3.4}$$

That is to say, (3.3) is the definition of (3.4).

The fact that intrigues some people is the appearance of the $R/6$ term because if we started out to quantize the system by means of a path integral then there would be no *a priori* reason to include it, and it would then reappear as a potential in the Schrödinger equation, assuming that we retraced the steps that led from (3.1) to (3.4) [14; p. 174]. One method of retracing the route from (3.1) to (3.4) is to avoid the Van Vleck WKB expression by deriving the Schrödinger equation directly from (3.4), in the form of (3.3), in the original way of Feynman [20]. Such a calculation is mentioned by DeWitt [11] and attributed to J. L. Anderson. More recently it has been detailed by Cheng [6].

By assuming that (3.4) means (3.3) we have automatically fixed the ordering of H by the built-in covariance, as mentioned earlier. If however we look upon (3.4) as a functional integral *still to be defined*, then the ordering ambiguities are present precisely because the integral is not uniquely defined. S is a stochastic integral (of a more general kind than that discussed by Ito [28]), different definitions of which will yield different Hamiltonian orderings [41].

What the author wishes to do here is simply to go through this reverse calculation from (3.4) to (3.1), assuming (3.3). This is a purely technical exercise of course but it is to be hoped that we shall learn a little about path-integrals on the way.

In order not to simply repeat earlier calculations (e.g. [6]), albeit in a more efficient way, the $A_\alpha \dot{q}^\alpha$ term will be included, as in (3.2). We can talk about this as an electromagnetic interaction. Later the discussion will be generalized to include the non-Abelian case.

If $r = 3$ we will just get the ordinary non-relativistic Schrödinger equation with a magnetic field interaction ($\mathbf{H} = \mathbf{\nabla} \times \mathbf{A}$). If $r = 4$ and $g_{\alpha\beta}$ has signature -2 we will obtain the Klein–Gordon equation, with an electromagnetic interaction, in Fock's fifth-parameter form ([21], [45]). Incidentally in this case, if the $R/6$ term is left out of the exponent in (3.4), it provides, in the massless Klein–Gordon equation, just the term necessary for conformal invariance [50]. It should be possible to prove the conformal invariance directly from the functional integral but note that this appears to work only for $r = 4$. The term needed for conformal invariance is $(r-2)/4(r-1)R$, while the functional integral gives $R/6$, independently of r. Also note that even though R may be constant it still has an effect in the relativistic case, since it adds into the (mass)^2 term.

We now proceed to the actual calculation and begin in standard fashion with the basic equation

$$\psi(q'', t'') = \int_M \langle q'', t'' | q', t' \rangle \psi(q', t') \, \mathrm{d}q', \tag{3.5}$$

where $\mathrm{d}q'$ stands for the invariant volume element on M, and then follow

Covariant Schrödinger equations

Feynman to write

$$-i\frac{\partial \psi(q'', t'')}{\partial t''} = \lim(i\varepsilon)^{-1}\left[\int \langle q'', t''|q', t''-\varepsilon\rangle \psi(q', t'')\, dq' - \psi(q'', t'')\right]. \quad (3.6)$$

You will recognize that this is precisely the way Kolmogorov derived the diffusion equations (see e.g. [25], [33; p. 60]). We seek to expand $\psi(q', t'')$ about q'' in a power series and then use moments, or averages, of these powers against $\langle q'', t''|q', t''-\varepsilon\rangle$.

First we consider the quantity

$$\lim_{\varepsilon \to 0} (i\varepsilon)^{-1} \int \langle q'', t''|q', t''-\varepsilon\rangle\, dq', \quad (3.7)$$

which we shall need at some point. In diffusion theory this quantity equals $\lim (i\varepsilon)^{-1}$ (total probability = 1), but not so here. The Feynman postulate, (3.4) and (3.3) gives the effective limiting short-time propagator

$$\lim_{\varepsilon \to 0} \langle q'', t''|q', t''-\varepsilon\rangle = \lim_{\varepsilon \to 0} \mathcal{N}(\varepsilon) \exp[iS(q'', t''|q', t''-\varepsilon) + \tfrac{1}{6}R''\varepsilon] \quad (3.8)$$

and even if $g_{\alpha\beta}$ were constant the $A_\alpha \dot{q}^\alpha$ term would contribute a potential $\sim A_\alpha A^\alpha$ to expression (3.7). The derivation of Schrödinger's equation for a charged particle in a magnetic field is a standard one [22; Problem 4.2].

If done in a direct way we obtain Schrödinger's equation in the 'expanded' form

$$i\dot{\psi} = -\tfrac{1}{2}[\nabla^2 \psi - 2i\mathbf{A}\cdot\nabla\psi - i(\nabla\cdot\mathbf{A})\psi - \mathbf{A}^2\psi].$$

What we should like is to produce the gauge-covariant Hamiltonian, $-\tfrac{1}{2}(\nabla - i\mathbf{A})^2$, all in one piece, and this is achieved by writing everything in terms of gauge-standardized quantities in the manner of DeWitt [12] and Mandelstam [35]. As will be shown this allows us to 'cancel' the $A_\alpha \dot{q}^\alpha$ term in the action.

The regular parametric curves $Z^\mu(q, q_0, \lambda)$

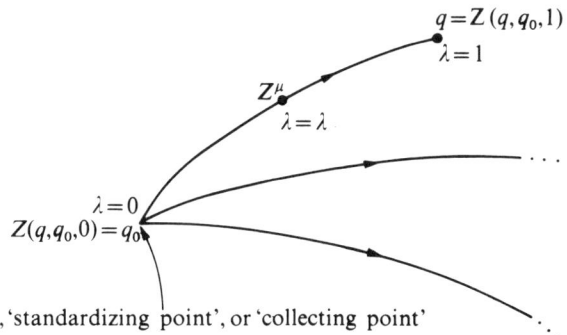

'Origin', 'standardizing point', or 'collecting point'

Covariant Schrödinger equations

We introduce a set $z^\mu = z^\mu(q, q_0, \lambda)$ of regular, parametric curves in M passing through the origin, or standardizing point q_0, such that each point q lies on only one curve. We consider this set of curves as fixed so that specifying q specifies the curve on which it sits. The functions z^μ satisfy

$$z^\mu(q, q_0, 1) = q^\mu, \qquad z^\mu(q, q_0, 0) = q_0^\mu,$$

$$\frac{\partial z^\mu(q, q_0, 1)}{\partial q^\nu} = \delta^\mu_\nu, \qquad \frac{\partial z^\mu(q, q_0, 0)}{\partial q^\nu} = 0. \qquad (3.9)$$

Then

$$\int_{P'q_0}^{q'} A_\alpha \, dq^\alpha = \int_0^1 A_\alpha(z) \frac{\partial z^\alpha(q', q_0, \lambda)}{\partial \lambda} \, d\lambda, \qquad (3.10)$$

where P' is the curve on which q' lies. (Sometimes it is convenient to use one form and sometimes the other.)

Gauge-standardized wave functions, $\Psi(q, t)$ are defined by the (gauge) transformation

$$\Psi(q, t) = \exp\left[-i \int_0^1 A_\alpha(z) \frac{\partial z^\alpha(q, q_0, \lambda)}{\partial \lambda} \, d\lambda\right] \psi(q, t) \equiv U\psi. \qquad (3.11)$$

If the ψ are subjected to a gauge transformation $\psi(q) \to \exp[i\Lambda(q)]\psi(q)$ then the Ψ transform uniformly, or in standardized fashion, $\Psi(q) \to \exp[i\Lambda(q_0)]\Psi(q)$. The gauge-standardized derivative of Ψ is given by $\mathcal{D}_\alpha \Psi$, with

$$\mathcal{D}_\alpha = U(\partial_\alpha - iA_\alpha)U^{-1} = \left[\partial_\alpha - i \int_0^1 F_{\beta\gamma}(z) \frac{\partial Z^\beta}{\partial q^\alpha} \frac{\partial Z^\gamma}{\partial \lambda} \, d\lambda\right],$$

and satisfies, by virtue of (3.11),

$$\mathcal{D}_\alpha \mathcal{D}_\beta \ldots \mathcal{D}_\gamma \Psi = U(\partial_\alpha - iA_\alpha)(\partial_\beta - iA_\beta) \ldots (\partial_\gamma - iA_\gamma)\psi \qquad (3.12)$$

(cf. [56]).

We can now write the gauge-standardized version of the propagation equation (3.5), namely,

$$\Psi(q'', t'') = \int_M \{q'', t''|q', t'\} \Psi(q', t') \, dq', \qquad (3.13)$$

where the gauge invariant propagator is given by

$$\{q'', t''|q', t'\} = \exp\left(i \int_{q''}^{q'} A_\alpha \, dq^\alpha\right) \langle q'', t''|q', t'\rangle. \qquad (3.14)$$

The path of integration from q'' to q' is backwards along $P'' = z^\alpha(q'', q_0, \lambda)$, to q_0 and then out again to q' along $P' = z^\alpha(q', q_0, \lambda)$.

Covariant Schrödinger equations

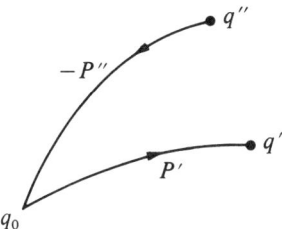

Next, the path-integral representation (3.4) is employed and the $\int_{q'}^{q''} A_\alpha \, dq^\alpha$ in this is combined with the phase term in (3.14) to yield a loop-integral, thus

$$\{q'', t''|q', t'\} = \mathcal{N} \int \exp\left[iS_0 + i \int_{t'}^{t''} \frac{1}{6} R \, dt + \oint_L A_\alpha \, dq^\alpha \right] \mathcal{D}[q], \quad (3.15)$$

where

$$S_0(q'', t''|q', t') = \frac{1}{2} \int_{t'}^{t''} g_{\alpha\beta} \dot{q}^\alpha \dot{q}^\beta \, dt,$$

and the loop L consists of the regular-curve part $P'-P''$ and the stochastic-path part $q(t)$ from q' to q'', $L = P' - P'' + [q(t)]$.

It is tempting to use, without further thought, Stokes's theorem to convert the loop-integral into a surface one,

$$\oint_L A_\alpha \, dq^\alpha = \int_\sigma F_{\alpha\beta} \, dS^{\alpha\beta}, \quad L = \partial\sigma. \quad (3.16)$$

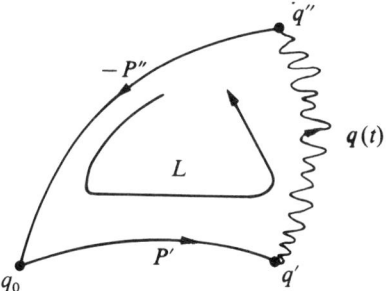

However, this is valid only when the implied partial integration is allowed. In our case where the integration over the stochastic paths is defined by (3.3) the conversion *is* justified and, in fact, is guaranteed automatically because gauge-invariance is maintained throughout the limiting process leading to the infinitely fine mesh.

Covariant Schrödinger equations

Speaking generally, the integral $\int_{q'}^{q''} A_\alpha \, dq^\alpha$ in (3.4) is a stochastic integral, and definition (3.3) tells us its meaning. In fact it is easy to show that it equals a 'symmetrized' stochastic integral† which, for convenience, can be taken to be the Stratonovich integral (or the average of the backwards and forwards Ito integrals) to which the ordinary rules of partial integration, etc. apply. Thus, as pointed out by McLaughlin and Schulman [39], gauge-covariance, in the quantum-mechanical sense, would make a definite choice of summation procedure in (3.4), (if it had not been already defined by (3.3)). Another way of saying this is that gauge-covariance fixes a choice of ordering, just like general covariance. This is obvious of course in terms of the Schrödinger equation, but here we see it emerging from the functional integral.

Returning to (3.15), with (3.16), we write down the differential form of (3.13),

$$-i\frac{\partial \Psi(q'', t'')}{\partial t''} = \lim_{\varepsilon \to 0} (i\varepsilon)^{-1}\left[\int \{q'', t''|q', t''-\varepsilon\}\Psi(q', t'')\, dq' - \Psi(q'', t'')\right] \quad (3.17)$$

with

$$\lim_{\varepsilon \to 0} \{q'', t''|q', t'\} = \lim_{\varepsilon \to 0} \mathcal{N}(\varepsilon) \exp\left(\frac{i}{2\varepsilon} s^2(q'', q') + \frac{i}{6} R''\varepsilon + i\int_{\sigma_\varepsilon} F_{\alpha\beta}\, dS^{\alpha\beta}\right), \quad (3.18)$$

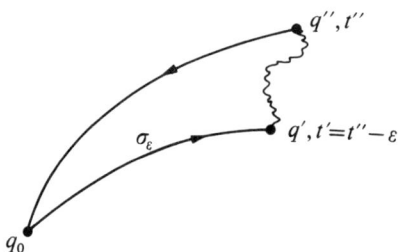

where $s(q'', q')$ is the geodesic distance between q'' and q'.

Now consider the quantity

$$\lim_{\varepsilon \to 0} (i\varepsilon)^{-1} \int \{q'', t''|q', t''-\varepsilon\}\, dq' = \lim_{\varepsilon \to 0} (i\varepsilon)^{-1} - V''. \quad (3.19)$$

† This point is discussed further in the appendix at the end of this chapter.

Covariant Schrödinger equations

Normal coordinates

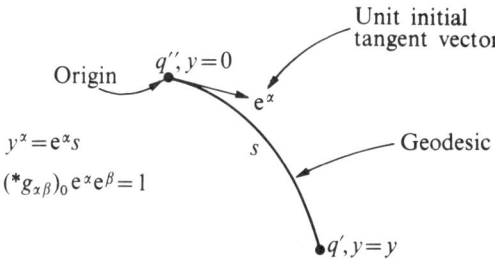

We want the integral over $F_{\alpha\beta}$ in (3.18) not to contribute to V. If the last exponential in (3.18) is expanded in a power series and the terms integrated against the first exponential, which is basically a Gaussian, we see that we shall want $\int_{\sigma_e} F_{\alpha\beta}\,dS^{\alpha\beta}$ to tend to zero faster than $(q''-q')^2$ as q' tends to q''. This can be done by choosing the paths z^μ appropriately. In order that the integral go to zero faster than $(q''-q')$ we must choose q'' as the origin q_0. For it to go faster than $(q''-q')^2$ the curves must tend to geodesics sufficiently fast at q' and q''. Obviously the best (only?) choice is just the set of geodesics through q''.

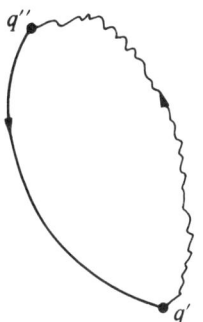

In this case we can forget the $\int F_{\alpha\beta}\,dS^{\alpha\beta}$ in (3.18) as it will contribute only terms of too high an order in ε to count.

We must now deal with the curved-space complications. Without more ado we introduce normal coordinates y^α (e.g. [67]) with origin at $q'' = q_0$. In terms of these, $s^2(q'', q') = (*g_{\alpha\beta})_0 y^\alpha y^\beta$ and the expression for dq' is

$$dq' = {}^*g^{\frac{1}{2}} \prod_\alpha dy^\alpha = (*g^{\frac{1}{2}})_0(1 - \tfrac{1}{6}(*R_{\beta\gamma})_0 y^\beta y^\gamma + \ldots) \prod_\alpha dy^\alpha. \quad (3.20)$$

Covariant Schrödinger equations

These forms are substituted into (3.18) which is then used to find the limit (3.19). The only integrals needed are that of a Gaussian and its second moment. It is easily checked that $V = 0$, the $R/6$ coming from the expansion of the exponential cancelling an $R/6$ from the average of $*g^i$. Also we find the standard normalization $\mathcal{N}(\varepsilon) = (2\pi i \varepsilon)^{-r/2}$.

The next step is to expand $\Psi(q', t'')$ about q'' in an ordinary Taylor series,

$$\Psi(q', t'') = *\Psi(y, t'') = *\Psi(0, t'') + y^\alpha \, \partial_\alpha \, *\Psi(y, t'')|_{y=0} + \tfrac{1}{2} y^\alpha y^\beta \partial_\alpha \partial_\beta \, *\Psi|_{y=0}$$
$$+ \ldots \tag{3.21}$$

$$(\Psi(q'', t'') = *\Psi(0, t'')),$$

substitute into (3.17) and again perform the Gaussian integrals. By virtue of (3.19) (with $V=0$) the first term in (3.21) cancels and we are left with Gaussian moments of the first order and upwards. The odd moments vanish and the fourth, and higher, ones contribute zero as ε vanishes. Thus only the first term of (3.20) and the third term of (3.21) survive to give Schrödinger's equation in the form

$$-i \frac{\partial \, *\Psi(y, t'')}{\partial t''}\bigg|_{y=0} = \tfrac{1}{2}(*g^{\alpha\beta})_0 \partial_\alpha \partial_\beta \, *\Psi(y, t'')\bigg|_{y=0}. \tag{3.22}$$

This is an invariant equation and is probably as good a way of writing the wave equation as any; however, let us restore the general covariance and gauge dependence by removing the restriction to normal coordinates and also the dependence on the paths z^α, i.e. let us return to the wave function $\psi(q, t)$. First we deal with the path-dependence by using eqn (3.12), in normal coordinates ($*z^\alpha(y, 0, \lambda) = y^\alpha \lambda$)

$$\left[\partial_\alpha - iy^\gamma \int_0^1 *F_{\alpha\gamma}(y\lambda)\lambda \, d\lambda\right]\left[\partial_\beta - iy^\delta \int_0^1 *F_{\beta\delta}(y\lambda)\lambda \, d\lambda\right] *\Psi(y, t'')$$

$$= \exp\left[-iy^\delta \int_0^1 *A_\delta(y\lambda) \, d\lambda\right][\partial_\alpha - i*A_\alpha(y)][\partial_\beta - i*A_\beta(y)] *\psi(y, t'').$$

If the limit $y \to 0$ is taken, and the equation rearranged we find

$$\partial_\alpha \partial_\beta \, *\Psi(y, t'')|_{y=0} = [(\partial_\alpha - i*A_\alpha)(\partial_\beta - i*A_\beta) *\psi(y, t'')$$
$$+ \tfrac{1}{2}i *F_{\alpha\beta}(y) *\psi(y, t'')]_{y=0} \tag{3.23}$$

(the right-hand side is symmetrical, as required).

To introduce covariance we make use of the concept of tensor extension due to Veblen [67] (see also [58], [54], [55]). By definition, the value of the extension of a tensor **T**, at any point q, is equal to the ordinary derivatives of the tensor in normal coordinates with q as origin, evaluated at this origin,

$$T^{\cdots}_{\|\alpha\cdots\beta}(q) \equiv \partial_\alpha \ldots \partial_\beta \, *T^{\cdots}(y)|_{y=0}.$$

It differs from the multiple covariant derivative by terms involving the curvature, in general. If T is a scalar its extensions are just the symmetrized covariant derivatives,

$$\psi_{\|\alpha\ldots\beta} = \psi_{\|(\alpha\ldots\|\beta)}.$$

From these results and eqn (3.23) we see that (3.22) gives the answer we expected from the start,

$$i\frac{\partial \psi(q'', t'')}{\partial t''} = -\tfrac{1}{2}(\nabla^\alpha - iA''^\alpha)(\nabla_\alpha - iA''_\alpha)\psi(q'', t'') \quad (3.24)$$

$$(\nabla_\alpha \psi \equiv \psi_{\|\alpha}).$$

The advantage of the present calculation is that it is covariant (or, in the form presented here, invariant) throughout, in contrast to the calculation of Cheng [6] or to the corrresponding derivations of the diffusion equation on Riemannian spaces ([18], [32], [70]).

If the manifold possesses some symmetry then this might force us naturally into a covariant calculation. Thus if M is a group manifold, we would use the generators to develop the expansions needed and the Hamiltonian would appear immediately as the Casimir operator. A calculation in which this occurs is that of McKean [36] on Brownian motion on SO(3). He uses the Ito stochastic differential technique and the exponential mapping (see also [38], [24], [18]).

A fact worthy of notice is that the WKB short-time propagator is exact for quantum mechanics on the manifolds of Lie groups† ([59], [60], [15], [16]). This was known (but not expressed in such terms) for Brownian motion on SO(3) ([5], see also [23]). The rather simple case of SO(2) was done by De Haas–Lorentz in 1914 (see [51]); Pauli [49] and Born and Ludwig [5] have discussed this example using quantum-mechanical language (see also [46] and references therein, and [33]).

It should be possible to derive this result directly from the functional integral, rather like the calculations for the free particle and harmonic oscillator. In fact these systems are obtained by taking an appropriate contraction of a Lie group.

It is of some interest to discuss the equations of motion for fields other than scalar ones, like ψ. Ito [29] has briefly considered the Brownian diffusion of tensor fields on Riemannian manifolds, but there are obscurities in his argument and, according to him, the computation of the diffusion generator (Hamiltonian) is complicated. A systematic discussion is possible and will be given here. Specifically, ψ becomes a spinor-valued quantity. If desired we can then return to tensors by, roughly speaking, projecting the appropriate spinor onto local Cartesian coordinate frames.

† Adding the contributions from all geodesics yields the propagator as the derivative of a theta function ([16], [19]).

Spinors are introduced pointwise into the manifold M by considering the representations of the homogeneous group of automorphisms $SO(p, r-p)$ of the (pseudo-)Euclidean tangent spaces (cf. [17], and references therein), More mathematical descriptions are available (e.g. [34]). The theory is required to be covariant under general coordinate transformations, ψ being a scalar with respect to these, and also under the family of position-dependent 'spin-basis', or similarity, transformations $\psi \to S\psi$ (where $S \in GL(l, C)$, l being the dimension of ψ). This latter covariance entails the introduction of a spinor covariant derivative, $\nabla_\alpha \psi = \partial_\alpha \psi + \Gamma_\alpha \psi$, where the matrix connection Γ_α is, or can be, determined from the requirement that the covariant derivatives of the fundamental connecting quantities between M and spinor space (e.g. γ^μ for Dirac theory) vanish; just as the Christoffel connection is determined from the vanishing of the covariant derivatives of the metric.

The minimal 'Riemannian' form Γ_α can take is a linear combination of the generators of $SO(p, r-p)$. The maintenance of such a form restricts the spin-basis transformations S to be elements of $SO(p, r-p)$ and the formalism is then the basic gauge, or Yang–Mills, one.

The ∇_α possess a generalized Ricci identity,

$$\nabla_{[\alpha} \nabla_{\beta]} \psi = \tfrac{1}{2} \Gamma_{\alpha\beta} \psi,$$

where $\Gamma_{\alpha\beta}$ is the spinor curvature,

$$\tfrac{1}{2} \Gamma_{\alpha\beta} = \partial_{[\alpha} \Gamma_{\beta]} + \Gamma_{[\alpha} \Gamma_{\beta]}.$$

Under a spin-basis transformation $\psi \to S\psi$, Γ_α transforms as

$$\Gamma_\alpha \to S \Gamma_\alpha S^{-1} - \partial_\alpha S \cdot S^{-1}, \tag{3.25}$$

so that

$$\nabla_\alpha \psi \to S \nabla_\alpha \psi.$$

We must now consider the propagation equation of the spinor. By analogy with the electromagnetic case we simply postulate the sum-over-paths form for the matrix propagator,

$$\langle q'', t'' | q', t' \rangle = \int \exp\left\{ iS_0 + \frac{i}{6} \int_{t'}^{t''} R \, dt \right\} T\left\{ \exp \int_{q'}^{q''} \Gamma_\alpha \, dq^\alpha \right\} \mathcal{D}[q], \tag{3.26}$$

where we have introduced the time-ordering symbol T, with respect to the parameter labelling the position along the stochastic path $q(t)$, which stands

Covariant Schrödinger equations

for a product integral (see [58], p. 340 and references therein; also see [3], [26], and the standard works on gauge theories). The time-ordered factor in (3.26) is the parallel propagator of ψ from q' to q'' along the Brownian path $q(t)$. We expect that ψ, propagated by (3.26), will satisfy the spinor generalization of (3.1). The way this works out is a simple extension of the previous calculation to the non-Abelian case. Since we have already made an investment in the path-dependent formalism it would be a pity not to use it, and we can be suitably brief.

Again the basic idea is to effectively cancel, to the necessary order in ε, the $\int \Gamma_\alpha \, dq^\alpha$ in (3.26) by an appropriately chosen path-dependent spinor wave function Ψ,

$$\Psi(q, t) = T \exp\left[\int_0^1 \Gamma_\alpha(z) \frac{\partial z^\alpha(q, q_0, \lambda)}{\partial \lambda} \, d\lambda\right] \psi(q, t) \equiv U\psi. \qquad (3.27)$$

Then, using the same notation,

$$\{q'', t''|q', t'\} = \mathcal{N} \int \exp i\left(S_0 + \frac{1}{6}\int_{t'}^{t''} R \, dt\right) T \exp\left(\oint \Gamma_\alpha \, dq^\alpha\right) \mathcal{D}[q],$$

which, after a non-Abelian version of Stokes's theorem (cf. [57]) has been used, becomes

$$\mathcal{N} \int \exp\left(iS_0 + \frac{i}{6}\int_{t'}^{t''} R \, dt\right) \exp\left[\int_\sigma U(\bar{q})\Gamma_{\alpha\beta}(\bar{q})U^{-1}(\bar{q}) \, d\bar{S}^{\alpha\beta}\right] \mathcal{D}[q].$$

The $U(\bar{q})$ terms transport the field $\Gamma_{\alpha\beta}(\bar{q})$ from the point $\bar{q}$ on the surface σ to the 'collecting point' q_0, where all contributions from σ are added to yield an object transforming in a standardized manner.

Precisely as before, if $q_0 = q''$ and the paths z^α are the set of geodesics through q'', the integral over σ will not contribute to the various limits as $t' \to t''$ and everything will go through as before. Thus eqns (3.17), (3.18), (3.19), with $v = 0$, and eqn (3.20) will still be valid, except that $\int F \cdot dS$ has become $\int U\Gamma \cdot U^{-1} \, dS$, and so Schrödinger's equation is the same as (3.22). The problems arise when we translate *$\Psi(y, t)$ back to $\psi(q, t)$. To do this we shall need the extensions of a spinor, i.e. we want the analogue of eqn (3.12). This is derived as follows.

The relation (3.27) between Ψ and ψ is a spin-basis transformation and so we can use eqn (3.25) which implies that

$$\mathcal{D}_\alpha \mathcal{D}_\beta \ldots \mathcal{D}_\gamma \Psi(q, t) = U(q)(\partial_\alpha + \Gamma_\alpha)(\partial_\beta + \Gamma_\beta) \ldots (\partial_\gamma + \Gamma_\gamma)\psi(q, t), \qquad (3.28)$$

where $\mathcal{D}_\alpha = \partial_\alpha + G_\alpha$ and G_α is determined by

$$G_\alpha = U\Gamma_\alpha U^{-1} - \partial_\alpha U \cdot U^{-1}.$$

Although the evaluation of this expression is straightforward it may be worthwhile giving it here.

W have to find $\partial_\alpha U$ and for this we use the formula for the variation of a product integral,

$$\delta\left\{\text{T}\exp\left[\int_0^1 \Gamma_\alpha(z)\frac{\partial z^\alpha}{\partial \lambda}\,d\lambda\right]\right\} = \int_0^1 U(z)\delta\left(\Gamma_\alpha\frac{\partial z^\alpha}{\partial \lambda}\right)U^{-1}(z)\,d\lambda \cdot U(q),$$

(e.g. [58; p. 340]), where $U(z)$ is the parallel propagator from $q_0 = q''$, to the point $z^\alpha(q, q_0, \lambda)$.

If we write $\delta(\ ) = \partial_\beta(\ )\,dq^\beta$ then we can find $\partial_\beta U(q)$ by expanding the $\partial_\beta\bigl(\Gamma_\alpha(\partial z^\alpha/\partial \lambda)\bigr)$ term and integrating the resulting $\Gamma_\alpha\partial_\beta(\partial z^\alpha/\partial \lambda)$ by parts. In doing this we shall have to use the other basic property of product integrals,

$$\frac{\partial U(z(q, q_0, \lambda))}{\partial \lambda} = U\Gamma_\alpha\frac{\partial z^\alpha}{\partial \lambda}.$$

Thus G_α is given by

$$G_\alpha = -\int_0^1 U(z)\Gamma_{\beta\gamma}(z)U^{-1}(z)\frac{\partial z^\beta}{\partial q^\alpha}\frac{\partial z^\gamma}{\partial \lambda}\,d\lambda.$$

Then, applying (3.28) to $\mathscr{D}_\alpha\mathscr{D}_\beta\Psi$ in normal coordinates and choosing the paths z^α to be geodesics through $q_o = q''$, we find

$$\partial_\alpha\partial_\beta {}^*\Psi(y, t)\big|_{y=0} = \nabla''_{(\alpha}\nabla''_{\beta)}\psi'',$$

whence Schrödinger's equation

$$i\frac{\partial \psi}{\partial t} = -\tfrac{1}{2}\nabla^\alpha\nabla_\alpha\psi.$$

These last steps follow after remarking that $\partial_\alpha\psi$ and $\Gamma_\alpha\psi$ transform as vectors under general coordinate transformations.

3.2. Solving the equations

The 'derivation' of the covariant Schrödinger equation just given is, in some ways, a sterile exercise, although we do learn a little about ordering and the stochastic nature of the path-integral. Much more interesting is the solution of these equations. However, only a very few words can be said on this topic.

In general terms the explicit construction of the Green's function goes back to the classic work of Hadamard, and it would be out of place here to follow through all later developments. We shall simply refer to the extensive article by Riesz [53].

More explicitly we note that the Van Vleck WKB expression, referred to earlier, plays the role of a parametrix, in the sense of Hilbert and Levi, and that Levi's integral equation expansion for the Green's function is a perturbation expansion in the quantum potential (e.g. [9]) Q, which here equals $(\Delta_2{}^*g^{-\frac{1}{4}}/2{}^*g^{-\frac{1}{4}})$ in normal coordinates (or, better, $\frac{1}{2}(\Delta_2\rho^{-\frac{1}{2}}/\rho^{-\frac{1}{2}})$, where ρ is Ruse's invariant [15]. $K = K_w - iKQK_w$.

This expansion has been used by McKean and Singer [37] to obtain expressions for the first few coefficients in the expansion for the 'partition function',

$$Z = \int_M \langle q, t|q, o\rangle \, dq = \int_M \langle q, t|q, 0\rangle_{WKB}(a_0 + a_1 it + a_2(it)^2 + \ldots) \, dq,$$

ignoring boundary effects. These coefficients can also be found, somewhat more easily, from the recursion formulae they satisfy (e.g. [43] in the manner discussed by DeWitt [13], see also [2]). (If M is a Lie group, $Q = \frac{1}{12}R$, is constant and the series is just $\exp(i\frac{1}{12}Rt)$. The WKB expression is then exact, as mentioned earlier.)

The calculation rapidly becomes tedious, and it may be that the expansion can be obtained from the functional-integral representation since this is geared to the classical limit. Corrections to the quasi-classical partition function have been derived in this way (e.g. [22]; see also [25]). In any event it seems to the author that there should be a rapid method of evaluating these coefficients. For partial confirmation we refer to the elegant expansion of Walker [68] for Ruse's invariant—essentially the Van Vleck determinant.

The extension of the parametrix method to higher spin has been performed, effectively, by Milgram and Rosenbloom [42], who discuss the propagation of p-forms (see also [37]). A different mode of description is employed by DeWitt [13], for spin $\frac{1}{2}$ and gauge fields. The corresponding WKB expression is the scalar one, $D^{\frac{1}{2}}\exp(iS_0)$ multiplied by the geodesic parallel propagator. This follows from our eqn (3.26).

We see, in these equations, plenty of opportunity for the development of the formalism.

Two final references concerning the covariant eqn (3.1) are to Balian and Bloch [1] and Minakshisundaram [44].

3.3. Conclusion

At some stage in the 'process' of quantizing a classical system, information, 'gauge invariance', 'general covariance', etc., has to be fed into the calculation. In the path-integral formulation this tells us what the functional integral means. It seems to me that there is no great advantage in this. One might as well feed the information directly into writing down a Schrödinger equation.

Of course a functional formulation has great elegance, 'it looks nice', and it may be useful in discussions involving invariance or constraints, but it does not solve any fundamental problems. The ordering question is still there, for example. Still, it would be interesting to know whether there should be an $R/6$ team in eqn (3.1). Only experiment can decide.

3.4. Appendix
A class of stochastic integrals

Consider a multidimensional diffusion process $q(t) = \{\ldots q^\alpha(t) \ldots\}$ described by the drift 'vector' $a^\alpha(q, t)$ and the local variance matrix $b^{\alpha\beta}(q, t)$, then we *define* a stochastic integral

$$\int \Phi_\alpha(q(t), t)\, dq^\alpha = \lim_{\substack{\varepsilon \to 0 \\ N \to \infty}} \sum_{i=0}^{N-1} \Phi_\alpha(q_{i+1}, q_i, t_i)(q_{i+1}^\alpha - q_i^\alpha), \qquad (3.A1)$$

where $q_i = q(t_i)$ and $\varepsilon = \max(t_{i+1} - t_i)$. The only restriction on $\Phi_\alpha(q_{i+1}, q_i, t_i)$ is that it should tend to $\Phi_\alpha(q_i, t_i)$ as t_{i+1} tends to t_i.

In order to express this in a more convenient form we introduce a matrix function $\mathbf{F}(\xi, \eta, t)$ of two 'vectors' $\xi = \{\xi^\alpha\}$ and $\eta = \{\eta_\alpha\}$ such that

$$\mathbf{F}(0, \eta, t) = \mathbf{1} = \mathbf{F}(\xi, 0, t). \qquad (3.A2)$$

Then we can write the matrix equation

$$\boldsymbol{\Phi}(q'', q', t) = \mathbf{F}(q'' - q', -i\bar{\partial}, t)\boldsymbol{\phi}(\bar{q}, t), \qquad (3.A3)$$

where

$$\bar{q} = (q' + q'')/2, \qquad \bar{\partial} = \partial/\partial\bar{q}$$

(cf. [40]). (The factors of i here, and later, are included because we are more interested in quantum mechanics than diffusion.)

For example, $\mathbf{F}(\xi, \eta, t) = \exp[(i/2)\xi \cdot \eta]\mathbf{1}$ gives the Ito forwards integral while $\exp[-(i/2)\xi \cdot \eta]\mathbf{1}$ gives the backward one. The 'simplest' choice, $\mathbf{F} = \mathbf{1}$, yields the Stratonovich midpoint integral [63].

More generally than Stratonovich we shall call the stochastic integral (3.A1), a 'symmetrized' integral if $\Phi_\alpha(q'', q', t) = \Phi_\alpha(q', q'', t)$. This implies the condition on $\mathbf{F}$,

$$\mathbf{F}(\xi, \eta, t) = \mathbf{F}(-\xi, \eta, t). \qquad (3.A4)$$

In order to emphasize the form (3.A2), the differential in the definition (3.A1) will be written $d^F q$. If the integral is a symmetrized one we write $d^s q$.

We now wish to show that all symmetrized integrals are equivalent. Thus the Stratonovich integral is essentially unique.

First, the difference between two different stochastic integrals is calculated,

$$\int \Phi_\alpha (d^{F_1}q^\alpha - d^{F_2}q^\alpha) = \lim_{\substack{\varepsilon \to 0 \\ N \to \infty}} \sum_{i=0}^{N-1} [F_1(q_{i+1} - q_i, -i\bar{\partial}_{i+1,i}, t_i)$$
$$- F_2(q_{i+1} - q_i, -i\bar{\partial}_{i+1,i}, t_i)]_{\alpha\beta} \Phi_\beta(\bar{q}_{i+1,i}, t_i)(q^\alpha_{i+1} - q^\alpha_i).$$
(3.A5)

The Fs are expanded in $(q_{i+1} - q_i)$ and the standard theorems on the averages of powers of $(q_{i+1} - q_i)$ used (cf. [63, pp. 45–50]). Without going into details, powers higher than the second can be ignored on the right-hand side of (3.A5) and the product $(q^\alpha_{i+1} - q^\alpha_i)(q^\beta_{i+1} - q^\beta_i)$ can be replaced by $\varepsilon i b^{\alpha\beta}(q_i)$ in the limit. Thus, in the expansion of $F_1(\xi, \eta, t) - F_2(\xi, \eta, t)$, all we need is the first term in ξ, and we arrive at the general formula

$$\int \Phi_\alpha(d^{F_1}q^\alpha - d^{F_2}q^\alpha) = i \int \left\{ \left[\frac{\partial}{\partial \xi^\gamma} (F_1(\xi, -i\partial, t) \right. \right.$$
$$\left. \left. - F_2(\xi, -i\partial, t))_{\alpha\beta} \right]_{\xi=0} \Phi_\beta(q, t) \right\} b^{\gamma\alpha}(q, t) \, dt$$
(3.A6)

independently of the drift a^α.

The condition for the two integrals to be equal is then that

$$\left. \frac{\partial \mathbf{F}_1(\xi, \eta, t)}{\partial \xi^\gamma} \right|_{\xi=0} = \left. \frac{\partial \mathbf{F}_2(\xi, \eta, t)}{\partial \xi^\gamma} \right|_{\xi=0}$$
(3.A7)

which is valid, in particular, if both $\mathbf{F}_1$ and $\mathbf{F}_2$ give symmetrized integrals, for then by (3.A4) both sides of (3.A7) are zero. This is the result we sought to prove.

We can check that (3.A6) gives the correct answer for the relation between the Ito and Stratonovich integrals by putting $\mathbf{F}_1 = \exp[(i/2)\xi \cdot \eta]\mathbf{1}$ and $\mathbf{F}_2 = \mathbf{1}$, whence

$$\int \Phi_\alpha(d^S q^\alpha - d^I q^\alpha) = \tfrac{1}{2} \int \partial_\beta \Phi_\alpha b^{\alpha\beta} \, dt,$$

which is just Stratonovich's theorem 2.2 [63].

Symmetrized stochastic integrals possess time symmetry and their uniqueness property corresponds precisely to the (well-known) unique Hermitian ordering of $pf(q)$, i.e. $\tfrac{1}{2}[\hat{p}, f(\hat{q})]_+$, in quantum mechanics.

The Hermitian condition is built into the lattice definition (3.3) of the functional integral (3.4), and so the stochastic integral $\int A_\alpha \, dq^\alpha$ occurring in (3.4) cannot be of the symmetrized type. We can think of this situation in the following way. Definition (3.3) tells us that the expression

(3.A1) for the stochastic integral $\int A_\alpha \, dq^\alpha$ is not adequate. Rather, we should consider it as a first term in some sort of expansion. In terms of (3.A1) it is easy to show that the value of $\int A_\alpha \, dq^\alpha$ in (3.4) should be taken as

$$\int A_\alpha \, dq^\alpha = \int A_\alpha \, d^F q^\alpha - i \int \left[\frac{\partial F_{\alpha\beta}(\xi, -i\partial)}{\partial \xi^\gamma} \bigg|_{\xi=0} A_\beta \right] b^{\alpha\gamma} \, dt, \quad (3.A8)$$

in order to agree with (3.3). The second integral on the right-hand side of (3.A8) is the effect of the higher-order terms. It is zero if $d^F q = d^S q$, from (3.A4).

Things are not so simple for the S_0 term in (3.4), i.e. for the stochastic integral $\int g_{\alpha\beta} \dot{q}^\alpha \dot{q}^\beta \, dt$. Here, Hermiticity or time-symmetry, is not sufficient, and appeal must be made to some other condition (e.g. covariance) to fix an ordering. Again this corresponds exactly to the fact that there is no unique Hermitian ordering of $p^2 f(q)$. This question is considered elsewhere ([40], [41], and references therein). It seems that no definition of the stochastic integral, within the class considered, will yield Δ_2. Higher-order terms are always non-zero, although it may be possible to give a *covariant* definition of the stochastic integral for which this would not then be true.

Acknowledgements

I would like to thank Professor J. Eells for drawing my attention to the work of V. K. Patodi [47] on the eigenforms of the Laplacian.

There is a useful list of references on the more mathematical aspects of diffusion theory on manifolds in Chapter 5.

On the more physical side, in addition to the works of Perrin and others cited earlier, we should mention the review article by Valiev and Ivanov [65] on rotational Brownian motion.

References

1. BALIAN, R. and BLOCH, G. *Ann. Phys.* **64**, 271 (1971).
2. BERGER, M. *C.r. hebd. Séanc. Sci., Paris* **263**, 13 (1966).
3. BERGMANN, P. G. *Handb. Phys.* **4**, 213 (1962).
4. BOCHNER, S. *International Congress of Mathematics, Cambridge*, p. 189 (1950).
5. BORN, M. and LUDWIG, G. *Z. Phys.* **150**, 106 (1958).
6. CHENG, K. S. *J. Math. Phys.* **13**, 1723 (1972).
7. CHERNOFF, P. R. *J. funct. Anal.* **12**, 401 (1973).
8. COHEN, L. *J. Math. Phys.* **11**, 3296 (1970).
9. DE BROGLIE, L. *Wave Mechanics*. Methuen, London (1930).
10. DEWITT, B. S. *Phys. Rev.* **85**, 653 (1952).
11. DEWITT, B. S. *Rev. mod. Phys.* **29**, 377 (1951).
12. DEWITT, B. S. *Phys. Rev.* **125**, 2189 (1962).
13. DEWITT, B. S. *Relativity, groups and topology*. Gordon and Breach, New York (1964).

14. DEWITT, CÉCILE M. *Ann. Inst. H. Poincaré N.S.* **11**, 152 (1969).
15. DOWKER, J. S. *J. Phys. A.* **3**, 451 (1970).
16. DOWKER, J. S. *Ann. Phys.* **62**, 361 (1971).
17. DOWKER, J. S. and DOWKER, Y. P. *Proc. phys. Soc.* **87**, 65 (1966).
18. DYNKIN, D. *Trans. Am. math. Soc.* **72**, 203 (1968).
19. ESKIN, L. D. *In Memoriam, N. G. Cebotarev*, p. 113, Kazan University (1964).
20. FEYNMAN, R. P. *Rev. mod. Phys.* **20**, 327 (1948).
21. FEYNMAN, R. P. *Phys. Rev.* **80**, 440 (1950).
22. FEYNMAN, R. P. and HIBBS, A. R. *Quantum mechanics and path integrals.* McGraw-Hill, New York (1965).
23. FURRY, W. H. *Phys. Rev.* **107**, 7 (1957).
24. GANGOLLI, R. *Z. Wahrschein.* **2**, 209 (1964).
25. GELFAND, I. M. and YAGLOM, A. M. *J. Math. Phys.* **1**, 48 (1960).
26. HAMILTON, J. F. and SCHULMAN, L. S. *J. Math. Phys.* **12**, 161 (1971).
27. HELGASON, S. *Differential geometry and symmetric spaces.* Academic Press, New York (1962).
28. ITO, K. *Nagoya Math. J.* **1**, 35 (1950).
29. ITO, K. *International Congress of Mathematicians, Stockholm*, p. 537 (1963).
30. JENSEN, H. and KOPPE, H. *Ann. Phys.* **63**, 586 (1971).
31. KLAUDER, J. R. *Ann. Phys.* **11**, 123 (1960).
32. KOLMOGOROV, A. N. *Math. Annln* **113**, 766 (1937).
33. LÉVY, P. *Processus stochastique et mouvement Brownian.* Gauthier–Villers, Paris (1948).
34. LICHNEROWICZ, A. In *Relativity, groups and topology* (eds. C. M. DeWitt and B. S. DeWitt). Gordon and Breach, New York (1964).
35. MANDELSTAM, S. *Ann. Phys.* **19**, 1 (1962).
36. MCKEAN, H. P. *Mem. Coll. Sci. Kyoto Univ.* **33**, 25 (1960).
37. MCKEAN, H. P. and SINGER, I. M. *J. diff. Geom.* **1**, 43 (1967).
38. MCKEAN, H. P. *Stochastic integrals.* Academic Press, New York (1969).
39. MCLAUGHLIN, D. W. and SCHULMAN, L. S. *J. Math. Phys.* **12**, 2520 (1971).
40. MAYES, I. W. and DOWKER, J. S. *Proc. R. Soc.* **A327**, 131 (1972).
41. MAYES, I. W. and DOWKER, J. S. *J. Math. Phys.* **14**, 434 (1973).
42. MILGRAM, A. N. and ROSENBLOOM, P. C. *Proc. natn. Acad. Sci. U.S.A.* **37**, 180 (1951).
43. MINAKSHISUNDARAM, S. and PLEIJEL, A. *Can. J. Math.* **1**, 242 (1949).
44. MINAKSHISUNDARAM, S. *J. ind. Math. Soc.* **17**, 158 (1953).
45. MORETTE, C. *Phys. Rev.* **81**, 848 (1951).
46. NORCLIFFE, A. *Case Stud. atom. Phys.* **4**, No. 1 (1973).
47. PATODI, V. K. *J. diff. Geom.* **5**, 233 (1971).
48. PAULI, W. *Handb. Phys.* **24**, 120 (1933).
49. PAULI, W. *Feldquantisierung.* Lecture Notes, Zurich (1951).
50. PENROSE, R. *Proc. R. Soc.* **A284**, 159 (1965).
51. PERRIN, F. *Ann. École. Norm. Supl.* **45**, 1 (1928).
52. PODOLSKY, B. *Phys. Rev.* **32**, 812 (1928).
53. REISZ, M. *Acta. math.* **81**, 1 (1949).
54. RUSE, H. S. *Proc. Lond. math. Soc.* **32**, 87 (1931).
55. RUSE, H. S. *Proc. Lond. math. Soc.* **33**, 66 (1932).
56. SCHIFF, L. I. *Phys. Rev.* **160**, 1257 (1967).
57. SCHLESINGER, L. *Math. Annln.* **99**, 413 (1927).
58. SCHOUTEN, J. A. *Ricci-Calculus.* Springer–Verlag, Berlin (1954).
59. SCHULMAN, L. S. *Phys Rev.* **176**, 1558 (1968).

60. SCHULMAN, L. S. Thesis, Princeton University (1967).
61. SCHRÖDINGER, E. *Annln Phys.* **79**, 748 (1926).
62. SOMMERFELD, A. *Wellenmechanik.* Ungar, New York (1928).
63. STRATONOVICH, R. L. *Conditional Markov processes and their application to the theory of optimal control.* Elsevier, New York (1968).
64. TESTA, F. J. *J. Math. Phys.* **12**, 1471 (1971).
65. VALIEV, K. A. and IVANOV, E. N. *Sov. Phys. Usv.* **16**, 1 (1973).
66. VAN VLECK, J. H. *Proc. natn. Acad. Sci. U.S.A.* **14**, 178 (1928).
67. VEBLEN, O. *Invariants of quadratic differential forms.* Cambridge University Press (1927).
68. WALKER, A. G. *Proc. Edin. math. Soc.* **7**, 16 (1942).
69. WELKER, H. *Math. Annln* **113**, 304 (1936).
70. YOSIDA, K. *Pacif. J. Math.* **2**, 263 (1952).

4. Functional problems in the theory of polymers

S. F. EDWARDS

4.1. Introduction

POLYMER molecules often take up random-flight shapes. The physics of such molecules, singly or in aggregate, is dominated by long-range correlations along their length, and the short-range structure can all be absorbed into a few parameters. Thus if we have a particular chemical composition repeated N times, and the chemical units, the monomers, have a series of positions available to them relative to their neighbours, then if one point on the first monomer is labelled R_1 and the corresponding point on the nth is R_n

$$(R_1 - R_n)^2 \propto n = nl^2, \quad \text{say}. \tag{4.1}$$

This defines l and says that, however complicated the monomer is, the large-scale probability distribution of configurations is the same as that of a chain of hinged rods each of length l, or indeed of any other system behaving as a random flight in the large. As the basic physics is dominated by these long-range effects one can turn the label n in R_n into a continuous variable s and talk of $R(s)$, which will then lead to Weiner measure, and the familiar Kac theorems are available. In the same way the interaction between monomers will be the usual compound of short-range repulsions and long-range attraction and can be compressed into a point interaction, a pseudopotential,

$$v \iint \delta(R(s) - R(s')) \, ds \, ds'.$$

When v is positive the chain repels itself. At a certain temperature $T = \theta$ there will be a balance between repulsion and attraction and $v = 0$. (For $T < \theta$ the chain will soon collapse, mathematically and physically, so this case will not be pursued.) There now opens up the vista of an entire branch of science, a vital one in both pure and applied science, completely dominated theoretically by functional integration. Surprisingly, this has been largely neglected by theoretical physicists, with the result that the experimental situation lacks the precision available elsewhere. Both theory and experiment are exacting, but a great storehouse of fascinating problems await study. In this chapter we shall look at three of them, but we can be assured that there are many more.

4.2. A problem from dilute solution theory

If a solution is dilute, a polymer molecule can be studied in isolation, and a suitable solvent and temperature found so that it behaves, on average, as a true random flight. If the solvent is sheared, the polymer molecules add to the viscosity. It was noted many years ago that the key to this problem lies in the fact that the distortion of the flow pattern around one bit of the molecule influences the flow pattern around another bit and leads to an increment in the viscosity which varies like $\rho M^{\frac{1}{2}}$ where ρ is the density and M the molecular weight. One always gets the ρ of course, but the M represents the length (on average) of the molecules. A detailed account of the history of this problem is given in a text by Yamakawa [4]. However, the treatments so far have not, to the author's mind, laid bare the statistical problem which has to be solved, and which Karl Freed and the present author now claim to have put in a clear form.

The viscosity increment can be expressed as

$$\left\langle \frac{\rho}{L} \int_0^L \int_0^L \zeta(ss')\,ds\,ds' \right\rangle = \frac{\delta v}{v}, \tag{4.2}$$

where L represents the molecular weight expressed in terms of Nl, l being the effective step-length and N the number of step-lengths in the molecule. The equation for ζ is

$$\int_0^L \frac{1}{|R(s)-R(s'')|}\zeta(s'',s')\,ds'' = |R(s)-R(s')|^2 \tag{4.3}$$

and $\langle \zeta \rangle$ means take the average over a random-flight ensemble.

If we redefine

$$s \to \tilde{s}L$$

$$R \to \tilde{R}\left(\frac{L}{l}\right)^{\frac{1}{2}}$$

then the problem becomes

$$\int_0^1 \frac{1}{|\tilde{R}(\tilde{s})-\tilde{R}(\tilde{s}'')|}\zeta(\tilde{s}'',\tilde{s}')\,d\tilde{s}'' = \left(\frac{L}{l^3}\right)^{\frac{1}{2}}|\tilde{R}(\tilde{s})-\tilde{R}(\tilde{s}')|^2, \tag{4.4}$$

i.e.

$$\zeta \propto L^{\frac{1}{2}} \tag{4.5}$$

so that

$$\frac{\delta v}{v} \propto \rho L^{\frac{1}{2}}, \tag{4.6}$$

the result long since obtained. But what is the coefficient and how quite generally does one go about solving equations like (4.4)?

Is it possible to give a bound to ζ and hence to $\delta \nu$? If one averages both left-hand and right-hand sides of the equation, this makes ζ no longer a functional of R, i.e.

$$\int_0^L \frac{1}{\sqrt{(l|s-s''|)}} \zeta(s''s') \, ds'' = \left(\frac{\pi}{6}\right)^{\frac{1}{2}} l|s-s'|, \tag{4.7}$$

a straightforward equation which has probably been studied in far more detail than it deserves, because there are large fluctuations in both sides of the equation—fluctuations which cannot be quantified by a coupling constant. At higher densities, we have shown that molecules interfere with each other and effectively screen the interaction, so that

$$\frac{1}{|R(s)-R(s')|} \quad \text{becomes modified to} \quad \frac{\exp[-\rho l|R(s)-R(s')|]}{|R(s)-R(s')|}, \tag{4.8}$$

and a coupling constant does now appear which permits a proper solution. This suggests that a large molecule will produce a micro-environment and screen itself, so that a better approximation than (4.3) might be

$$\int_0^L \frac{\exp[-|s-s''|/(Ll)^{\frac{1}{2}}]}{\sqrt{(l|s-s''|)}} \zeta(s''s') \, ds'' = l|s-s'|, \tag{4.9}$$

and such can indeed be derived by a physical argument. Again, however, there are fluctuations ignored; one just argues physically that (4.9) is better than (4.7) as an approximation to (4.3). What would be very useful would be a method of putting a bound on ζ, which the author thinks would have to be an upper bound, but all attempts to do this have given bounds which are too crude to be useful.

4.3. Rubber elasticity

Consider an assembly of chains which are chemically cross-linked so that if S_α^n is a point α on the nth chain, and S_β^m similarly:

$$R^{(n)}(S_\alpha^{(n)}) = R^{(m)}(s_\beta^{(m)}).$$

This gives a model for a rubber, the n_c chemical links, which we can label $(n\alpha; m\beta)$, holding the assembly of chains together. Since it is known experimentally that the free energy is approximately proportional to temperature one may safely ignore all forces in the first approximation and relate all properties of the rubber to the entropy, i.e. to the integration over all configurations of the system. But the links are permanent—they do not slip. This means that the average over the links must be made in the free energy, *not* in its exponential, which would correspond to slipping links which cannot form a solid. Thus it is wrong to say

$$\exp\left[-\frac{F}{kT}\right] = \int \exp\left[-\frac{3}{(2l)}\sum_n \int R'^{(n)^2}(s)\,ds\right]\int\int\prod_{\substack{nm\\\alpha\beta}}\delta[R^{(n)}(s_\alpha^n)-R^{(m)}(s_\beta^m)]$$
$$\prod\delta R^{(n)}\,ds_\alpha^n\,ds_\beta^m. \tag{4.10}$$

$$\underset{m \text{ terms}}{\longleftrightarrow}$$

Rather, one must write

$$\exp\left[-\frac{F(\text{all}\{n\alpha;\,m\beta\})}{kT}\right] = \int \exp\left[-\frac{3}{(2l)}\sum_n \int R'^{(n)^2}\,ds\right]\prod\delta[R^n(s_\alpha^n)$$
$$-R^m(s_\beta^m)]\prod\delta R^n, \tag{4.11}$$

and finally

$$F = \int F(\{n\alpha;\,m\beta\})\prod ds_\alpha^n\,ds_\beta^m. \tag{4.12}$$

These integrals look, and indeed are, obscure, and the literature of this subject has chosen particular models and solved them outside a general formalism. For example, if one assumes that when a rubber is distorted the links are embedded in a fixed way in the matrix of rubber and distort affinely, then one can say: let two links be a distance R apart. The piece of chain between them has an entropy $s(R)$; i.e.

$$F = \sum_{\text{pieces}} kTS(R) = F(R). \tag{4.13}$$

Now deform the body so that $R \to AR$, e.g.

$$A = \begin{bmatrix} \lambda_1 & 0 & 0 \\ 0 & \lambda_2 & 0 \\ 0 & 0 & \lambda_3 \end{bmatrix}$$

for a body stretched by $\lambda_1, \lambda_2, \lambda_3$ in the x, y, z directions.
Then

$$F(\lambda_1, \lambda_2, \lambda_3) = \int F(AR)p(R)\,d^3R, \tag{4.14}$$

where $p(R)$ is the probability of finding cross-links a distance R apart in the first place. Which is fine of course providing the model is right, but we would like a basic formalism to see whether it is right. We have tried to do this by exploiting the identity

$$\ln x = \text{coefficient of } m \text{ in } x^m. \tag{4.15}$$

Now consider our box of rubber and m further boxes of rubber and label the $(m+1)$ chains $R_0\,R_1\ldots R_m$.

Functional problems in the theory of polymers

Let us now calculate a free energy for the $(m+1)$ systems $\mathcal{F}(m)$ defined by

$$\exp\left[-\frac{\mathcal{F}(m)}{kT}\right] = \int \prod_{a=0}^{m} \prod \delta R_a^{(n)} \exp\left(-\sum_n \sum_a \frac{3}{2l} \int R_a^{\prime(n)2} \, ds\right) \int \int \prod_a \delta[R_a^{(n)}(s) - R_a(s')] \, ds \, ds' \quad (4.16)$$

Note that all the $(m+1)$ systems can carry the *same* s label. At this point, since to have a rubber at all there must be many links per chain, we can consider the $R^{(n)}$ chains as one big chain, the equation for $\mathcal{F}$ is written

$$\exp\left[-\frac{\mathcal{F}(m)}{kT}\right] = \int \exp\left[-\sum_{a=0}^{m} \frac{3}{2l} \int R_a^{\prime 2}(s) \, ds\right] \\ \times \left\{\int \int \prod_a \delta[R_a(s) - R_a(s')] \, ds \, ds'\right\}^{n_c} \quad (4.17)$$

or

$$\oint \frac{d\mu n_c!}{\mu^{n_c+1}} \int \exp\left\{-\sum_0^m \frac{3}{2l} \int R_a^{\prime 2}(s) \, ds + \mu \int \int \prod_a \delta[R_a(s) - R_a(s')] \, ds \, ds'\right\} \quad (4.18)$$

Then

$$\mathcal{F}(m) = \mathcal{F}(O) + mF + O(m^2), \quad (4.19)$$

and the coefficient of m is the free energy. In formula (4.18) we have integrated over the 'incidence matrices' $\{n\alpha; m\beta\}$ and we have a complete formula at the cost of working in $3(m+1)$ dimensions. Now it turns out that one can actually use this formula. Everything is dominated by Gaussian integrals which can be done in an arbitrary number of dimensions, and the particular problem of elasticity amounts to integrating the R_0 over the *original* box, and the $R_1 \ldots R_m$ over the box distorted by $\lambda_1, \lambda_2, \lambda_3$, so that

$$F = F(\lambda_1, \lambda_2, \lambda_3) \quad (4.20)$$

Details are given in reference [2].

The author has recently found that the same method will give the cusp in the susceptibility of a spin glass, and it seems of general acceptability in disordered systems. But is it mathematically valid to integrate over a functional space of what amounts to a vector field in a continuous number of dimensions? What the author has done is to take m, an integer, then treat it as a continuous variable and let it tend to zero. The integrals all seem well behaved, but nevertheless there are strange features. The integral (4.18) is supposed to represent a solid, but (assuming for the moment there is no distortion $(\lambda_1 = \lambda_2 = \lambda_3 = 1)$ it is identical to an integral in a $(3m+3)$-dimensional space, so that, for example, if one looked at a correlation function for the density, one would expect this to be a function of $|\mathbf{r} - \mathbf{r}'|$, $\mathbf{r}$

being a vector in the $(3m+3)$-dimensional space. Such is clearly not a solid—compare with the ordinary space of three dimensions.

A correlation function of a homogeneous system will be

$$C(|r-r'|) = C(\sqrt{[(x-x')^2+(y-y')^2+(z-z')^2]}). \tag{4.21}$$

Now, if our integral represents a rubber it has to have correlation functions like

$C(|R_0-R_1|, |R_0-R_2|\ldots)$ which would be the analogues of

$$C([(x-x')-(y-y')]^2, [(x-x')-(z-z')]^2). \tag{4.22}$$

If one assumes the functions do have this form, one can indeed get an evaluation which fits a rubber, but my point is that this is introducing an order which is not displayed in (4.18) i.e. breaking a symmetry. Integral (4.18) then can represent both a solid rubbery state, and a chaotic state which we can identify with a dilutely cross-linked ungelled state. There must be a phase change implicit in (4.18). Do the manoeuvres of (4.16) to (4.19) survive the mathematical singularity of a phase change?

4.4. Entanglements of chains

In addition to the cross-links, the chains are entangled, and the presence of the cross-links means that the entanglements are permanent. To specify the topological relationship between a member of a network, an infinite number of invariants needs to be used, but at the crudest level one already finds fascinating problems. The simplest invariant is the angle swept out between chains familiar from Gauss and its application to current flow by Ampere. This gives that if $\mathbf{R}^{(1)}(s)$, $\mathbf{R}^{(2)}(s)$ are closed or infinite,

$$I_{12} = \int\int d\mathbf{R}^{(1)} \times d\mathbf{R}^{(2)} \nabla \frac{1}{|R^{(1)}-R^{(2)}|} = \text{constant} \tag{4.23}$$

and so is invariant under deformation in which one chain does not cross another. In two dimensions if $R^{(1)}$ lies in a plane, and $R^{(1)}$ is a (straight) line perpendicular to the plane, I_{12} represents the angle subtended by $R^{(1)}$ around the point at which $R^{(2)}$ cuts the plane. In three dimensions, for two rings, $I_{12} = \pm 4\pi$ or 0 according to whether the rings are linked, the sign being dependent on the sense in which $R(s)$ increases with s. So now in addition to the δ-functions from cross-links, one must put in a δ-function for the invariant between each pair, and strictly the higher invariant as well. Since the invariant I_{12} varies by $4\pi \times$ integer, the δ-function must be a Kronecker rather than a Dirac δ-function. The solution for the configurational statistics of random walks is soluble in simple two-dimensional cases, but in three dimensions is the same as quantum field theory [1]. The divergence difficulties of quantum field theory now emerge as the obvious difficulty that when two curves governed by a Wiener measure get close together they are infinitely entangled. Physically of course they are not because molecules have a finite l. Thus our original contention that only long distances matter seems to fail here. But on a closer look it is probably all right. For example, under shear, some tight little ball of entanglement will not deform affinely or just not deform at all. However, one obviously cannot just gloss over this difficulty and the author and his research student R. Deam have been able to make crude evaluations in detail of the effect of entanglement following earlier work confined to small deformations [2]. This amounts to approximating the Kronecker δ product

$$\prod_{a=1}^{m} \delta(I^{(0)} - I^{(a)}) \quad \text{by} \quad \exp\left[-\frac{\pi^2}{6} \sum_{a=1}^{m} (I^{(0)} - I^{(m)})^2\right]$$

and evaluating $(I^{(0)} - I^{(m)})^2$ just as one evaluates $\delta(R_0 - R_m)$ in the cross-link case. The mass of algebra is such that it will not be given here, but we shall merely observe that it makes a most interesting problem with many experimental consequences. A further development is to try to understand how an uncross-linked rubber can flow, as any melted plastic will. The entanglements are now dominant and determine the viscosity and visco-elasticity completely. Studies of these problems are not hopeless and progress is being made ([3], [5]) but the theories are not yet convincing. The fascinating thing is that, whereas other difficult areas of theoretical physics may be so esoteric as to defy visualization, the dynamics of polymer melts is basically that of pouring spaghetti. One has a quite clear picture of what is happening, but it is hard to put it into decent mathematics. Now of course one is dealing with $R(s, t)$ not just $R(s)$ or $R(t)$, and the extension of the normal theories to a two-variable path specification, though simple enough for Gaussian integrals, is not at all obvious for other cases. Thus to study

$$\exp\left\{-\beta \int \int ds\, dt [r\dot{R}(s, t) - kTR''(s, t)]^2\right\}$$

is quite straightforward and can be turned into a differential equation propagating either in t or in s, with the Eulerian variable $r(s)$ or $r(t)$ respectively. But such a simple form does not appear with hydronamics interactions or entanglements. So we are not likely to run out of puzzling and challenging problems for some while yet.

References

1. EDWARDS, S. F. *J. Phys.* A**6**, 15–28 (1968).
2. EDWARDS, S. F. In *Proceedings of ACS Symposium Chicago 1970*; published as *Polymer networks* (ed. C. Chompff). Plenum Press, New York (1971).
3. DE GENNES, P. G. *J. chem. Phys.* **55**, 572–9 (1971).
4. YAMAKAWA, H. *Modern theory of polymer solutions*. Harper–Row, New York (1971).
5. EDWARDS, S. F. and GRANT, J. W. V. *J. Phys.* A**6**, 1169–85, 1186–11 (1973).

5. Measures on infinite-dimensional manifolds

K. D. ELWORTHY

5.1. Introduction

THIS is intended as an introduction to three closely related approaches to obtaining measures on Banach manifolds from geometric structures imposed on the manifolds. The first is the abstract Wiener manifold approach, strongly based on the work of H-H. Kuo [19]; the other two depend on the theory of diffusions on manifolds with particular reference to work by Daletskii and Shnaiderman [5]. Most of the material is taken from the results of a joint investigation with J. Eells; since these results are treated in more detail elsewhere [9], [10], [11], [12], only brief summaries are given here.

5.2. Abstract Wiener manifolds

This approach gives a direct analogue in infinite dimensions of the finite-dimensional theory of smooth measures via scalar densities [23], with Lesbesgue measure replaced by a fixed Gaussian measure. More details are given in [10], [12].

Let E be a separable (real) Banach space, and γ a strictly positive Gaussian measure on E. Then γ determines a continuous linear injection $i: H \to E$, with dense image, of a separable Hilbert space H into E, such that if $p: E \to V$ is a continuous linear surjection onto an n-dimensional vector space V, and $B \subset V$ is a Borel set, then

$$\gamma(p^{-1}B) = (2\pi)^{-n/2} \int_B \exp\left(-\frac{|x|^2}{2}\right) dx$$

where V is furnished with the quotient inner product induced from the surjection on $p \circ i: H \to V$. The triple (i, H, E) is then an abstract Wiener space in the sense of L. Gross [15], [16] and γ is the corresponding abstract Wiener measure [7]. The measure γ is quasi-invariant under translation by the point x of E iff $x \in i(H)$.

A map $f: U \to E$ is called a W'-map if it has the form $f(x) = x + ij\alpha(x)$, where $\alpha: U \to E^*$ is C^r and $j: E^* \to H$ is obtained from the adjoint of i. The following theorem, that γ is quasi-invariant under W^1-diffeomorphisms is due to H-H. Kuo [19]: a stronger result is obtained by R. Ramer [30]:

THEOREM 5.1. *Let $\phi = l_E + ij\alpha: U \to V$ be a W^1-diffeomorphism between open subsets of E. Then $\phi^{-1}(\gamma|V)$ is equivalent to $\gamma|V$ and the Radon–Nikodym derivative*

$$g(\phi, -) \equiv \frac{d\phi^{-1}(\gamma)}{d\gamma}: U \to \mathbf{R}$$

Measures on infinite-dimensional manifolds 61

is given by

$$g(\phi, x) = |\det D\phi(x)| \exp\{\tfrac{1}{2}[-2\alpha(x)(x) - |j\alpha(x)|_H^2]\}.$$

(It follows from standard properties of abstract Wiener spaces that the derivative $D\phi(x)$ of ϕ can be assigned a well-defined determinant, at each point x of U.)

Following Kuo, define a C^r *abstract Wiener manifold* (AWM) over (i, H, E) to be a C^r manifold M modelled on E together with a maximal C^r atlas of the form $\{(U_i, \phi_i)\}, \phi_i: U_i \to E$, such that each $\phi_i \circ \phi_j^{-1}$ is a W^r-map on its domain. A *Wiener density* ξ on such a manifold is determined by a family $\{\xi_i\}_i$ of functions $\xi_i: U_i \to \mathbf{R}$ such that

$$g[\phi_j\phi_i^{-1}, \phi_i(x)]\xi_j(x) = \xi_i(x) \quad \text{if } x \in U_i \cap U_j.$$

The density is *positive* if each $\xi_i > 0$. From Kuo's theorem we see: *a C^1 abstract Wiener manifold structure over (i, H, E) on a Banach manifold M determines a measure class on M, the measures in the class corresponding to the positive Wiener densities of the structure.*

The most often-used densities in finite-dimensional differential geometry are those induced by Riemannian metrics. Because of the lack of translation invariance of our underlying Gaussian measure γ it turns out that in order to obtain analogues of these we need both a 'Riemannian metric' G on our abstract Wiener manifold M and a 'position field' Z on M. The position field is just a vector field on M which in each chart of the AWM is represented by a W^{r-1}-map, and the 'metric' G is a suitably chosen inner product defined on a dense subspace of each tangent space T_mM to M. In the chart (U_j, ϕ_j), G must be given by

$$\langle u, v \rangle_x = \langle G_x^j u, G_x^j v \rangle$$

for a linear W^∞-map $G_x^j: E \to E$, for each $x \in \phi_j(U_j)$, where $\langle \ , \ \rangle$ refers to the densely defined inner product induced on E by $i: H \to E$. In particular G gives each tangent space to M the structure of an abstract Wiener space. The pair (G, Z) then induces a positive Wiener density on M [12], and hence a positive measure on M. When the abstract Wiener manifold M is just E with its trivial structure we can take G to be $\langle \ , \ \rangle$ itself and define $Z(x) = x$. The measure corresponding to this (G, Z) is just the original measure γ.

There is an analogue of the classical divergence theorem for manifolds with boundary in the abstract Wiener manifold context [12], [18], and even a finite co-dimensional differential form theory [29]. Also data (G, Z) define a differential operator Δ on M which is an analogue of the Laplace–Beltrami operator; when $M = E$ as above this reduces to

$$\Delta f(x) = \operatorname{tr}(D^2 f(x) \circ (i \times i)) - Df(x)(x).$$

In general the theory of abstract Wiener manifolds and their measures seems to give a very straightforward extension of the finite-dimensional theory; the problem is that we have had to impose a strong additional structure on the underlying manifold M. Although for a wide class of Banach spaces E, e.g. all Hilbert spaces, every separable metrizable E manifold M admits an AWM structure, in many practical situations there is no natural one to choose. This is true, for example, in general, for manifolds of maps between finite-dimensional manifolds. However, every finite co-dimensional submanifold of E has an AWM structure induced on it by (i, H, E). It also has a canonical position field and metric, and hence it has a well-defined measure. Thus Brownian motion on $\mathbf{R}^n$ determines an AWM structure and measure on the manifold of all continuous paths $\sigma: (0, 1] \to \mathbf{R}^n$ with $\sigma(0) \in P$, $\sigma(1) \in Q$, where P and Q are submanifolds of $\mathbf{R}^n$.

5.3. Construction by stochastic differential equations

5.3.1. Let (i, H, E) be an abstract Wiener space as above. There is then a Brownian motion on E determined by the abstract Wiener space structure [17]. This can be considered as a Gaussian measure μ on the Banach space $\mathscr{C}_0(E)$ of all continuous paths $\sigma: [0, T] \to E$ with $\sigma(0) = 0$, for some interval $[0, T]$, together with the map

$$w: [0, T] \times \mathscr{C}_0(E) \to E$$

defined by $w(t, \sigma) = \sigma(t)$. The measure μ is constructed so that $w(t, -)(\mu) = \gamma_t$, where γ_t is the abstract Wiener measure on E with variance parameter t, defined by (i, H, E).

For a connected, separable C^3 Hilbert manifold M, let X be a C^2 section of the bundle Hom $(M \times E, TM)$ over M; so for each m in M we have a continuous linear map $X(m): E \to T_m M$. Also let A be a C^1, possibly time-dependent, vector field on M. Roughly speaking, we say that the process $x: [0, T] \times \mathscr{C}_0(E) \to M$ is a *solution of the stochastic differential equation* $dx = X\, dw + A\, dt$ if x is non-anticipating, has continuous sample paths, and in each chart of M satisfies the stochastic integral equation

$$x(t, \sigma) - x(s, \sigma) = \int_s^t X[x(u, \sigma)]\, dw(u, \sigma) + \int_s^t A[u, x(u, \sigma)]\, du +$$

$$+ \frac{1}{2} \int_s^t \operatorname{tr} DX[x(u, \sigma)] \circ X[x(u, \sigma)]\, du$$

(some care is needed to make this localization precise [11]. The trace involved in $\sum_j DX(x)(X(x)e_j, e_j)$, where the sum is taken over the image in E of an orthonormal base $\{e_j\}_{j=1}^{\infty}$ of H. The first integral is an Itô integral, or equivalently a McShane belated integral [25], [26]. In any case this way of interpreting stochastic differential equations in order to give them an invariant meaning comes from McShane [25], [26]. An essentially equivalent approach using the Fisk–Stratonovich integral is given in [4].

In general, solutions cannot be expected to exist for all time in $[0, T]$, unless M is compact. However, given an initial distribution $\tilde{x}(0): \mathscr{C}_0(E) \to M$ there is a unique maximal solution (up to the 'explosion time' ξ)

$$x:[0, \xi) \times \mathscr{C}_0(E) \to M \text{ with } x(0, \sigma) = \tilde{x}(0)(\sigma), \quad \sigma \in \mathscr{C}_0(E),$$

where $\xi: \mathscr{C}_0(E) \to (0, T]$. The solution is then a diffusion process on M associated to the time-dependent differential operator L^t on M which in a chart is given by

$$L^t(f)(x) = \tfrac{1}{2}\sum_j D^2 f(x)(Xe_j, Xe_j) + \tfrac{1}{2} Df(x)[\operatorname{tr} DX(x) \circ X(x)] + Df(x)[A(t, x)],$$

with $\{e_j\}_j$ as above.

Suppose now that we take such a solution x with initial distribution $\tilde{x}(0)(\sigma) = x_0$, all $\sigma \in \mathscr{C}_0(E)$, where x_0 is some point of M. Then:

(i) if $\xi(\sigma) = T$ for all σ we obtain a probability measure ν on the Banach manifold $\mathscr{C}_{x_0}(M)$ of all continuous paths $\rho: [0, T] \to M$ with $\rho(0) = x_0$ by setting $\nu = \phi(\mu)$, where $\phi: \mathscr{C}_0(e) \to \mathscr{C}_0(M)$ is defined, almost everywhere at least, by $\phi(\sigma)(t) = x(t, \sigma)$.

(ii) in any case we obtain Borel measures $p_t(x_0, -)$ on M, $0 \leq t < T$ by setting

$$p_t(x_0, B) = \mu\{\sigma: t < \xi(\sigma) \quad \text{and} \quad \sigma(t) \in B\}$$

for each Borel set B of M.

We thus obtain two more methods of constructing measures on certain manifolds. Method (i) is classical and applies only to path spaces. Method (ii) is a modification of the approach of Daletskii and Shnaiderman [5]. The restriction that M be a Hilbert manifold is for technical reasons, and is probably not essential. A discussion of stochastic differential equations on abstract Wiener manifolds is given by Kuo in [20], cf. Kuo–Pietch in [22]. The standard approach to diffusions on finite-dimensional manifolds is described in McKean's book [24].

5.3.2. Example (i) [10], [11], [12]

Let N be a connected n-dimensional Riemannian manifold without boundary and let $\pi: O(N) \to N$ be its orthonormal frame bundle. The Levi-Civita connection on N determines a horizontal subbundle of the tangent bundle $TO(N)$ together with a trivialization of that subbundle. Thus we obtain an $X: O(N) \times \mathbf{R}^n \to TO(N)$. The Brownian motion w on (i, H, E) with $H = E = \mathbf{R}^n$ is just n-dimensional Brownian motion. Let $x: [0, \xi) \times \mathscr{C}_0(\mathbf{R}^n) \to O(N)$ be the solution of the stochastic differential equation $dx =$

$X dw$ starting at the point u_0 of $O(N)$. Then $\pi \circ x: [0, \xi) \times \mathscr{C}_0(\mathbf{R}^n) \to N$ is a diffusion process on N with infinitesimal generator the Laplace–Beltrami operator of N [11]. When N is compact, and in some other situations [6] we can assume that $\pi \circ x$ is defined on the interval $[0, T]$, and so obtain a probability measure λ on $\mathscr{C}_{x_0}(N)$, where $x_0 = \pi(u_0)$. (Otherwise we obtain a measure on $\mathscr{C}_{x_0}(N^*)$, where N^* is the one-point compactification of N, cf. [27].)

The measure λ could be called the Wiener measure of N. An interesting point is that the manifold $\mathscr{C}_{x_0}(N)$ has the rudiments of the structure of an AWM modelled on the classical Wiener space $L_0^{2,1}(\mathbf{R}^n) \subseteq \mathscr{C}_0(\mathbf{R}^n)$, with the map $\mathscr{C}_0(\mathbf{R}^n) \to \mathscr{C}_0(N)$ induced by $\pi \circ x$ as a 'chart'. (The Hilbert space $L_0^{2,1}(\mathbf{R}^n)$ is given the inner product

$$\langle \sigma_1, \sigma_2 \rangle_{2,1} = \int_0^T \langle \sigma_1'(t), \sigma_2'(t) \rangle \, dt.$$

5.3.3. Example (ii)

Let M be a separable Lie group modelled on a Hilbert space. Let e denote its identity element, and set $E = T_e M$, the tangent space at e to M. There is then a left invariant trivialization of TM, $X: M \times E \to TM$, $X(m, v) = (L_m)_*(v)$. suppose $i: H \to E$ is an injective Hilbert–Schmidt map of a Hilbert space H into E. Then (i, H, E) is an abstract Wiener space, and so induces a Brownian motion on E as described above. The left invariance of X gives sufficient uniformity to ensure that the stochastic differential equation $dx = X\, dw$, $x(0) = e$, has a full solution

$$x: [0, T] \times \mathscr{C}_0(E) \to M.$$

Thus we obtain a map $\phi: \mathscr{C}_0(E) \to \mathscr{C}_e(M)$ and a probability measure ν on $\mathscr{C}_e(M)$ with $\phi(\mu) = \nu$. We also get the probability measures $p_t(e, -)$ on M, $0 \leq t \leq T$.

As a special case suppose that $i: H \to E$ is also the derivative map at e, $T_e h: T_e G \to T_e M$, of a smooth injective homomorphism $h: G \to M$, where G is a Lie group with a left invariant Riemannian metric. Let $\alpha: [0, T] \to G$ be a smooth path starting at e. Identifying $h[G]$ with G, define the process

$$y: [0, T] \times \mathscr{C}_0(E) \to M$$

by

$$y(t, \sigma) = \alpha(t) x(t, \sigma).$$

Then y satisfies the stochastic differential equation

$$dy = X\, dw + A\, dt, \qquad y(0) = e \ldots, \tag{5.1}$$

with A the time-dependent vector field on M:

$$A(t, m) = X(m)\mathrm{Ad}(m)^{-1}X[\alpha(t)]^{-1}\alpha'(t),$$

where $\mathrm{Ad}: G \to GL(E)$ denotes the adjoint representations of G on E.

On the other hand, define $T_\alpha : \mathscr{C}_0(E) \to \mathscr{C}_0(E)$ by

$$T_\alpha(\sigma) = \sigma + k_\alpha(\sigma)$$

where

$$k_\alpha(\sigma)(t) = \int_0^t \mathrm{Ad}[x(s, \sigma)]^{-1}X[\alpha(s)]^{-1}\alpha'(s)\,ds.$$

It is a straightforward verification that the process $(t, \sigma) \mapsto x[t, T_\alpha(\sigma)]$ satisfies (5.1), whence $y(t, \sigma) = x[t, T_\alpha(\sigma)]$. This shows that the measure $\alpha(\nu)$ on $\mathscr{C}_e(M)$ induced by y is also given by $\alpha(\nu) = \phi \circ T_\alpha(\mu)$. It follows that $\alpha(\nu) \approx \nu$ (i.e. ν is quasi-invariant under left multiplication in the group $\mathscr{C}_e(M)$ by the path α) iff $T_\alpha(\mu) \approx \mu$.

The measure μ is determined [1] by the abstract Wiener space $L_0^{2,1}(H) \to \mathscr{C}_0(E)$, and the transformation $T_\alpha : \mathscr{C}_0(E) \to \mathscr{C}_0(E)$ looks enticingly similar to a W^r transformation on this space (and even more like the transformations considered by Ramer [30]). This is especially so when G is normal in M, since then $\mathrm{Ad}(m)^{-1}X[\alpha(t)]^{-1}\alpha'(t)$ lies in H for all $m \in M$, $t \in [0, T]$. However, it seems as if the strongest results on quasi-invariance can be obtained by looking at eqn (5.1) directly and using probabilistic methods (see [5], [21], and [14]). Clearly quasi-invariance of ν under left multiplication by smooth paths in G implies quasi-invariance of each of the measures $p_t(e, -)$ on M under left multiplication by elements of G, at least when G is connected. Daletskii and Shnaiderman introduce this approach in [5], where they state their results without proof; some extra subtlety would seem to be needed in their arguments to cover the case when G is not normal in M.

As concrete examples of the above we could take M and G to be the Sobolev manifolds of maps [8],

$$M = H^r(K, A), \qquad G = H^s(K, A),$$

where A is a finite-dimensional Lie group and K is a compact n-dimensional smooth manifold, provided $2r > n$, $2(r-s) > n$. Replacing left invariance by right invariance the same technique could be applied to the diffeomorphism groups of K [8], [28]: $M = \mathscr{D}^r(K)$, $G = \mathscr{D}^s(K)$ for $r > n$, $2(r-s) > n$, although these are not Lie groups. Another special case is when M is finite-dimensional and $H = E$, $G = M$.

5.3.4. Stochastic differential equations are also used in a method, being developed by Baxendale, Eells, and Ramer, of constructing measures on spaces of maps between Riemannian manifolds.

5.4. General remarks

A suitable linear map $i: H \to E$ of a Hilbert space H into a Banach space E determines a Gaussian measure γ on E as described in § 5.2. Example (*ii*) shows that there is a corresponding result for suitable homomorphisms $h: G \to M$ of a Hilbert Lie group G with Riemannian metric; also the Hilbert manifold of paths $L^{2;1}_{x_0}(N)$ seems to play a similar role in Example (*i*) [11], [12]. An abstract Wiener manifold M contains a Hilbert manifold M_H, given by $M_H = \bigcup_j \phi_j^{-1}(H)$, which appears to determine the measure theoretic properties of M [29], although it has measure zero. One of the main problems in trying to make our constructions more useful is to understand more precisely the role of these Hilbert spaces and manifolds in the hope of being able to work solely in terms of them. Presumably this boils down to an attempt to construct some non-linear theory of weak distributions [31] or cylinder set measures.

There is also a converse problem. The Gaussian measure γ induces the map $i: H \to E$; given a measure μ on a manifold M what structure does it induce on M? For example, can we determine the group G and the homomorphism $h: G \to M$ of Example (*ii*) from the measure $p_t(e, -)$, for a particular $t \in [0, T]$? And can we determine $L^{2,1}_{x_0}(N)$ from the Wiener measure λ of N in Example (*i*)? Rather surprisingly any strictly positive measure μ on a smooth infinite-dimensional Banach manifold M does determine some non-trivial geometric structure on M [13]. If instead of a single measure of M we have a diffusion process on M the situation is particularly good, since the process determines its infinitesimal generator.

Bessaga's example [2], [3] of a diffeomorphism $f: H \to H - \{0\}$ on an infinite-dimensional Hilbert space H can be modified to show that there are limitations to the programme of trying to concentrate entirely on the underlying Hilbert spaces and manifolds even for Gaussian measures. Given (i, H, E) and γ as above, his technique yields a C^∞ diffeomorphism $\phi: E \to E$ whose restriction to $E - \{0\}$ is a W^∞-map but with $\phi(0) \notin i(H)$. From Kuo's theorem $\phi(\gamma) \approx \gamma$, however, ϕ does not preserve the subspace $i(H)$.

Acknowledgements

Besides the main sources quoted I am indebted to R. Ramer for emphasizing the role of diffusion processes and to K. Schmidt and R. Ramer for discussions about quasi-invariance (e.g. Example (*ii*), p.64).

References

1. BAXENDALE, P. Gaussian measures on function spaces, Thesis, University of Warwick (1973). Submitted to *Am. J. Math.*
2. BESSAGA, C. Every infinite dimensional Hilbert space is diffeomorphic with its unit sphere. *Bull. Acad. polon. Sci.* **14**, 27–31 (1966).

3. BURGHELEA, D. and KUIPER, N. H. Hilbert manifolds. *Ann. Math.* **90**, 379–417 (1969).
4. CLARK, J. M. C. An introduction to stochastic differential equations on manifolds. In (eds. D. Q. Mayne and R. W. Brockett). *Geometric methods in systems theory.* Reidel, Holland (1973).
5. DALETSKII, YU. L. and SHNAIDERMAN, YA. I. Diffusions and quasi-invariant measures on infinite-dimensional Lie groups. *Funktsional'nyi Analiz i Ego Prilozheniya* **3**, 88–90 (1969). English translation: *Funct. Anal. Appl.* **3**, 156–8 (1969).
6. DEBIARD, A. GAVEAU, B., and MAZET, E. Temps d'arret des diffusions riemanniennes. *C.r. Lebd. Séanc. Acad. Sci., Paris.* (To appear.)
7. DUDLEY, R. M., FELDMAN, J., and LECAM, L. On seminorms and probabilities, and abstract Wiener spaces. *Ann. Math.* **93**, 390–408 (1971).
8. MARSDEN, J. E., EBIN, D. G., and FISCHER, A. E. Diffeomorphism groups, hydrodynamics, and relativity. *Proceedings of the 13th Biennial Seminar of the Canadian Mathematical Congress, Halifax,* pp. 135–279 (1971).
9. EELLS, J. Integration on Banach manifolds. *Proceedings of the 13th Biennial Seminar of the Canadian Mathematical Congress, Halifax,* pp. 41–9 (1971).
10. EELLS, J. and ELWORTHY, K. D. Wiener integration on certain manifolds, 'Some problems in non-linear analysis'. *Centro Internazionale Matematico Estivo IV,* 67–94. Edizioni Cremonese, Rome (1971).
11. EELLS, J. and ELWORTHY, K. D. Stochastic differential geometry seminar notes, Mathematics Institute, University of Warwick, (In preparation.)
12. ELWORTHY, K. D. Gaussian measures on Banach spaces and manifolds. *Proceedings of the Summer Institute on global analysis, Trieste* (1972). (To appear.)
13. ELWORTHY, K. D. Differential invariants of measures on Banach spaces and manifolds. (In preparation.)
14. GIRSANOV, I. V. On transforming a certain class of stochastic processes by absolutely continuous substitution of measures. *Theory Prob. Appl.* **5**, 285–301 (1960).
15. GROSS, L. Measureable functions on Hilbert space. *Trans. Am. math. Soc.* **105**, 372–90 (1962).
16. GROSS, L. Abstract Wiener spaces. *Proceedings of the 5th Berkeley Symposium in mathematical statistics and probability,* pp. 31–42 (1965–6).
17. GROSS, L. Potential theory on Hilbert space. *J. funct. Anal.* **1**, 123–81 (1967).
18. GOODMAN, V. A divergence theorem for Hilbert spaces. *Trans. Am. Math. Soc.* **164**, 411–26 (1972).
19. KUO, H.-H. Integration theory in infinite dimensional manifolds. *Trans. Am. Math. Soc.* **159**, 57–78 (1971).
20. KUO, H.-H. Diffusion and Brownian motion on infinite dimensional manifolds. *Trans. Am. math. Soc.* **169**, 439–59 (1972).
21. KUO, H.-H. Absolute continuity of measures corresponding to diffusion processes in Banach space. *Ann. Prob.* **1**, 513–18 (1973).
22. KUO, H.-H. and PIETCH, M. A. Stochastic integrals and parabolic equations in Abstract Wiener Space. *Bull. Am. math. Soc.* **79**, 478–82 (1973).
23. LOOMIS, L. H. and STERNBERG, S. *Advanced calculus.* Addison-Wesley, New York (1968).
24. MCKEAN, H. P. *Stochastic integrals.* Academic Press, New York (1969).
25. MCSHANE, E. J. Stochastic differential equations and models of random processes, *Proceedings of the 6th Berkeley Symposium on mathematical Statistics and Probability, Vol. III,* pp. 263–94 (1970–1).

26. MCSHANE, E. J. Stochastic integration, *Proceedings of the Conference on vector and operator valued measures and applications*, Park City Utah (1972).
27. NELSON, E. An existence theorem for second order parabolic equations. *Trans. Am. math. Soc.* **88**, 414–29 (1958).
28. OMORI, H. Groups of diffeomorphisms and their subgroups. *Trans. Am. math. Soc.* **179**, 85–122 (1973).
29. RAMER, R. Integration of infinite dimensional manifolds. Thesis, University of Amsterdam (1974).
30. RAMER, R. On nonlinear transformations of Gaussian measures. *J. funct. Anal.* **15**, 166–87 (1974).
31. SEGAL, I. E. Algebraic integration theory. *Bull. Am. math. Soc.* **71**, 419–89 (1965).

6. The free Euclidean Proca and electromagnetic fields

L. GROSS

6.1. Introduction

THE recent successes of Euclidean quantum field theory in the work of Dobrushin and Minlos, Fröhlich, Glimm, Griffiths, Guerra, Jaffe, Nelson, Osterwalder, L. Rosen, Schrader, Simon, Spencer, and their collaborators (cf. [2], [3], [4], [5], [6], [7] and their bibliographies) for $P(\phi)_{2,3}$ theories and the applicability of statistical-mechanical methods, in particular, clearly demand that the higher-spin theories be investigated with a view to developing similar probabilistic techniques. It is to be hoped that these techniques can eventually be applied to fermion fields as well as boson fields in order to study experimentally observed interactions. In the present paper we describe Euclidean versions of the free Proca and electromagnetic fields in a form which seems to us likely to contain the appropriate notion of Euclidean locality for local quantum field theory. Our description is closely connected with Schwinger's source theory [11].

Let K be the one-particle Hilbert space of a relativistic integer-spin wave equation. K is then a space of positive energy solutions ϕ to the wave equation. To form the corresponding Euclidean theory we shall follow a three-step procedure.

Step 1

We identify the dual space K^* with a space of (vector-valued) functions f on $\mathbf{R}^3$ in the pairing

$$\langle f, \phi \rangle = \int_{\mathbf{R}^3} f(x)\phi(0, x) \, \mathrm{d}^3 x.$$

The square of the norm on K^* is essentially the fundamental quadratic form used by Schwinger on his source space, although K^* does not exactly coincide with his source space. K^* may be interpreted as a space of instantaneous sources at time zero.

Step 2

We describe a Hilbert space K_e of (generalized) vector-valued functions on $\mathbf{R}^4$ such that (a) the natural action of the Euclidean group on K_e is unitary and (b) the space of functions in K_e with support in the hyperplane $x_4 = 0$ coincides with K^*.

Step 3

We show that the symmetric Fock space over K is isomorphic to the 'time-zero' subspace $\mathscr{F}_0$ of $L^2(K_e, \text{normal distribution})$ in such a way that, among other things, $\exp(-tH_0) = E_0 T_t$, where T_t denotes time-translation in $L^2(K_e)$, E is the projection onto $\mathscr{F}_0$, and H_0 is the free field Hamiltonian on $\mathscr{F}_0$.

We shall carry out these three steps for the free Proca and electromagnetic fields. For the Klein–Gordon equation the one-particle space K is the Sobolev space $\mathscr{H}_{\frac{1}{2}}(\mathbf{R}^3)$ and its dual space K^* is $\mathscr{H}_{-\frac{1}{2}}(\mathbf{R}^3)$, which plays a by now familiar role in the transition to the Euclidean theory. The novelty will lie partly in the surprising way in which the not very Euclidean-looking K^* (the functions in K^* have only three components) emerges naturally from K_e (whose elements have four components) and partly in the related fact that gauge-invariance problems for the free electromagnetic field disappear in the Euclidean setting in spite of the continued use of the potentials. We also discuss the Markoff property (or absence thereof) for the Euclidean fields.

6.2. The Proca field

The classical Proca field of mass $m > 0$ is the field governed by the differential equations

$$\left(\Delta - \frac{\partial^2}{\partial t^2}\right)\phi_\alpha = m^2 \phi_\alpha, \qquad \alpha = 0, 1, 2, 3 \tag{6.1}$$

$$\sum_{\alpha=0}^{3} \frac{\partial \phi_\alpha}{\partial x_\alpha} = 0, \tag{6.2}$$

where $\{x_\alpha\} = (-t, \mathbf{x})$ and $\phi_\alpha(t, \mathbf{x})$ are real-valued functions and Δ denotes the three-dimensional Laplacian. However, in order to describe positive-energy solutions conveniently, we shall consider complex-valued solutions to (6.1) and (6.2). By choosing a suitable subspace of the space of all (distribution) solutions to (6.1) and (6.2) we obtain a Hilbert space which supports a unitary representation of the inhomogeneous Lorentz groups as follows.

Let $\mathscr{C}_m$ denote the mass hyperboloid given by

$$\mathscr{C}_m = \{k \in \mathbf{R}^4;\ k = (k_0, \mathbf{k}),\ k_0^2 = |\mathbf{k}|^2 + m^2,\ k_0 > 0\}.$$

$\mathscr{C}_m$ is of course invariant under the orthochronous homogeneous Lorentz group $\mathscr{L}_0$ and, as is known, the measure d^3k/k_0 on $\mathscr{C}_m$ is invariant under $\mathscr{L}_0$. In our description of this measure is implicit the choice of (k_1, k_2, k_3) as a global coordinate system on $\mathscr{C}_m$. Let T_k be the tangent plane to $\mathscr{C}_m$ at k for each k in $\mathscr{C}_m$. Clearly T_k is spacelike at each point k and the Lorentz-invariant inner product

$$k' \cdot k'' \equiv \mathbf{k}' \cdot \mathbf{k}'' - k_0' k_0''$$

is positive-definite on T_k. Let K denote the space of functions h on $\mathscr{C}_m$ to $\mathbf{R}^4 \otimes \mathbf{C}$ such that

$$h(k) \in T_k \otimes \mathbf{C} \quad \text{for all}\quad k \text{ in } \mathscr{C}_m \tag{6.3}$$

and

$$\|h\|^2 = \int_{\mathscr{C}_m} h(k) \cdot \overline{h(k)} \frac{d^3k}{k_0} < \infty. \tag{6.4}$$

Then, identifying functions equal almost everywhere on $\mathscr{C}_m$, K is a complex Hilbert space. For each h in K put

$$\phi(x) = (2\pi)^{-\frac{3}{2}} \int_{\mathscr{C}_m} h(k) e^{-ik.x} \frac{d^3k}{k_0}, \quad x \text{ in } \mathbf{R}^4. \tag{6.5}$$

Then ϕ is a complex four-vector valued distribution on $\mathbf{R}^4$, and it is verified readily that ϕ satisfies (6.1) and (6.2) in the distribution sense. Eqn (6.2) is a consequence of the tangentiality condition on h. The natural action of an orthochronous inhomogeneous Lorentz transformation, $Lx = Ax + b$, on ϕ, namely, $\phi \to A\phi L^{-1}$, induces a linear transformation on the Hilbert space K, which is easily verified to be unitary. Thus a unitary representation of the orthochronous inhomogeneous Lorentz group in K is obtained. What we have described is typical of the well-known way [1] in which a Lorentz-invariant partial differential equation leads to a unitary representation of the orthochronous Poincaré group. K is by definition the one-particle Hilbert space for the Proca field. Using (6.5) to identify the elements of K with solutions of (6.1) and (6.2) the norm of ϕ is by definition $\|\phi\| = \|h\|$. To complete this expository account of the one-particle space for the Proca field we mention that time-reversal acts in K as the anti-unitary operator induced in K by the map $\phi \to \phi'$, where $\phi'_\alpha(x, t) = -\overline{\phi^\alpha}(x, -t)$ ($\bar{a}$ = complex conjugate of a).

It is illuminating to describe the norm of an element h in K more directly in terms of the spatial part $\boldsymbol{\phi}$ of ϕ. For each k in $\mathscr{C}_m$ the tangent plane T_k is a direct sum of a two-dimensional subspace consisting of vectors parallel to the hyperplane $\{k_0 = 0\}$ (i.e. horizontal vectors—also called transverse vectors because they are perpendicular to $(0, \mathbf{k})$) and its one-dimensional orthogonal complement (the longitudinal vectors), each element of which has its spatial part parallel to $\mathbf{k}$. Thus for each function h in K we may write

$$h(k) = \mathbf{t}(k) + l(k), \tag{6.6}$$

where $\mathbf{t}(k)$ is a complex transverse (i.e. horizontal) vector in $T_k \otimes C$ and $l(k)$ has its spatial part parallel to $\mathbf{k}$. Then, since $\mathbf{t}(k)$ and $l(k)$ are Lorentz orthogonal for each k, we have $\|h\|^2 = \|\mathbf{t}\|^2 + \|l\|^2$. Writing ϕ^t and ϕ^l for the solutions to (6.1) and (6.2) obtained from t and l as ϕ was obtained from h in (6.5), we may write $\phi = \phi^t + \phi^l$.

For any finite-dimensional complex inner-product space V, any tempered distribution $f: \mathbf{R}^n \to V$ and any real number a, define the Sobolev norm

$$\|f\|_a^2 = \int_{\mathbf{R}^n} |\hat{f}(k)|^2 (m^2 + |k|^2)^a \, dR \tag{6.7}$$

whenever this is finite, where $\hat{f}(k) = (2\pi)^{-n/2} \int f(x) e^{ik \cdot x} dx$ and $k \cdot x = \sum_{j=1}^{n} k_j x_j$. We shall be interested primarily in the two cases $n = 3, 4$ and various values of a.

Put
$$\mu(k) = (m^2 + |k|^2)^{\frac{1}{2}}, \quad k \text{ in } \mathbf{R}^3.$$

With **t** and l as in (6.6) we see that
$$\|\mathbf{t}\|^2 = \int_{\mathbf{R}^3} |\mathbf{t}(k)/\mu(k)|^2 \mu(k) \, d^3k, \tag{6.8}$$

wherein we have identified a point in $\mathscr{C}_m$ with its projection in the hyperplane $k_0 = 0$ (so that we should really write $\mathbf{t}((\mathbf{k}, \mu(k)))$ since $(\mathbf{k}, \mu(k))$ is the point of $\mathscr{C}_m$ lying above **k**). Here and in the following we shall write k instead of **k** when it is clear from the context that k is in $\mathbf{R}^3$. Since $(2\pi)^{-\frac{3}{2}} \int_{\mathbf{R}^3} \phi'(x, 0) e^{ix \cdot \mathbf{k}} d^3k = \mathbf{t}(\mathbf{k})/\mu(\mathbf{k})$ by (6.5), we have

$$\|\phi'\| = \|\phi(\,.\,, 0)'\|_{\frac{1}{2}}. \tag{6.9}$$

On the other hand, writing $\mathbf{l}(k) = \alpha(k)\mathbf{k}$ for the spatial part of $l(k)$, where $\alpha(k)$ is a complex number, the tangentiality condition, $k \cdot l(k) = 0$, shows $k_0 l_0(k) = \alpha(k)|\mathbf{k}|^2$. Hence $l_0(k) = \alpha(k)|\mathbf{k}|^2/\mu(\mathbf{k})$. Thus the Lorentz inner product

$$l(k) \cdot \overline{l(k)} = |\mathbf{l}(k)|^2 - |l_0(k)|^2 = |\alpha(k)|^2|\mathbf{k}|^2 - |\alpha(k)|^2|\mathbf{k}|^4/\mu(\mathbf{k})^2$$
$$= m^2|\alpha(k)|^2|\mathbf{k}|^2/\mu(\mathbf{k})^2.$$

Hence
$$l(k) \cdot \overline{l(k)} = \frac{m^2 |\mathbf{l}(k)|^2}{\mu(\mathbf{k})^2}.$$

Therefore
$$\|l\|^2 = m^2 \int_{\mathbf{R}^3} \left| \frac{\mathbf{l}(k)}{\mu(k)} \right|^2 \mu(k)^{-1} \, dk. \tag{6.10}$$

It follows that the Lorentz norm of ϕ' can be expressed in terms of its spatial part $\phi'(x, 0)$ at time zero by a Sobolev norm;

$$\|\phi'\|^2 = m^2 \|\phi'(\,.\,, 0)\|^2_{-\frac{1}{2}}, \tag{6.11}$$

since $(2\pi)^{-3} \int_{\mathbf{R}^3} \phi'(0, x) e^{ix \cdot k} dx = \mathbf{l}(k)/\mu(\mathbf{k})$ by (6.5). Combining (6.9) and (6.11) we obtain

$$\|\phi\|^2 = \|\phi'(0, \,.\,)\|^2_{\frac{1}{2}} + m^2 \|\phi'(0, \,.\,)\|^2_{-\frac{1}{2}}. \tag{6.12}$$

We note that ϕ' has a zero fourth component while the fourth component of l and *a fortiori* the fourth component of ϕ' is uniquely determined by the spatial part of l and ϕ' respectively. Furthermore the decomposition of ϕ into a transverse and longitudinal part depends only on the spatial part of ϕ at time zero, i.e. on $\phi(0, x)$, since the three-dimensional Fourier transform of $\phi(0, x)$ is uniquely $\mathbf{t}(\mathbf{k})/\mu(\mathbf{k}) + \mathbf{l}(\mathbf{k})/\mu(\mathbf{k})$, where $\mathbf{t}(\mathbf{k})$ is orthogonal to $\mathbf{k}$ and $\mathbf{l}(\mathbf{k})$ is parallel to $\mathbf{k}$. Hence (6.12) shows that $\|\phi\|$ depends only on the spatial part of ϕ at time zero.

In the following we put $\boldsymbol{\nabla} \cdot f$ for $\sum_{j=1}^n \partial f_j/\partial x_j$ for any (generalized) function $f: \mathbf{R}^n \to \mathbf{R}^n \otimes C$.

THEOREM 6.1. *Let K^* be the set of all distributions $f: \mathbf{R}^3 \to \mathbf{R}^3 \otimes C$ such that*

$$\|f\|^2 = \|f\|^2_{-\frac{1}{2}} + m^{-2}\|\boldsymbol{\nabla} \cdot f\|^2_{-\frac{1}{2}} < \infty. \tag{6.13}$$

Then K^* is dual to the one-particle Proca field space K in the pairing

$$\langle f, \phi \rangle = \int_{\mathbf{R}^3} f(x) \cdot \phi(x, 0) \, dx. \tag{6.14}$$

Proof. Let f be in K^*. Decompose f into transverse and longitudinal parts by writing

$$\hat{f}(k) = u(k) + v(k), \tag{6.15}$$

where $u(k) \cdot k = 0$ and $v(k)$ is parallel to k. Then $(\boldsymbol{\nabla} \cdot \hat{f})(k) = -ik \cdot \hat{f}(k) = -ik \cdot v(k)$. Hence

$$\|f\|^2_{K^*} = \int_{\mathbf{R}^3} (|\hat{f}(k)|^2 + m^{-2}|-ik \cdot \hat{f}(k)|^2)\mu(k)^{-1} \, dk$$

$$= \int (|u(k)|^2 + |v(k)|^2 + m^{-2}|k \cdot v(k)|^2)\mu(k)^{-1} \, dk$$

$$= \int |u(k)|^2 \mu(k)^{-1} \, dk + \int (|v(k)|^2 + m^{-2}|k|^2|v(k)|^2)\mu(k)^{-1} \, dk.$$

Thus, since $1 + m^{-2}|k|^2 = m^{-2}\mu(k)^2$, we have

$$\|f\|^2_{K^*} = \int_{\mathbf{R}^3} |u(k)|^2 \mu(k)^{-1} \, dk + m^{-2} \int_{\mathbf{R}^3} |v(k)|^2 \mu(k) \, dk. \tag{6.16}$$

By the Plancherel theorem we have

$$\langle f, \phi \rangle = \int_{\mathbf{R}^3} (u(-k) + u(-k)) \cdot \phi(k, 0) \, dk$$

$$= \int \{u(-k) \cdot [\mathbf{t}(k)/\mu(k)] + v(-k) \cdot [\mathbf{l}(k)/\mu(k)]\} \, dk, \tag{6.17}$$

where we have used the notation preceding eqns (6.9) and (6.12). This computation is valid at least if f and $\phi(0, x)$ are, say, C^∞ functions with compact support. But comparison of (6.8), (6.10), and (6.12) with (6.16) and (6.17) shows that $|\langle f, \phi \rangle| \le \|f\|_{K^*} \|\phi\|$. Hence the right side of (6.14) has a unique jointly continuous bilinear extension to all of $K^* \times K$ which we take as the *definition* of $\langle f, \phi \rangle$. (This allows us to avoid technical discussion as to whether an element f in K^* is actually a function, so that $f(x) \cdot \phi(0, x)$ may be defined.) It is clear then that (6.17) is now valid for all f in K^* and ϕ in K. It is not hard to see that every continuous linear functional on K has the form (6.19) for some f in K^*. For, in fact, since K is a Hilbert space, every continuous linear functional β on K has the form $\beta(\phi) = (h, h')_K$ for some h' in K, where ϕ is given by (6.5). If $h'(k) = \mathbf{t}'(k) + l'(k)$ as in (6.6) then by polarization of (6.8) and (6.10):

$$(h, h')_K = (\mathbf{t}, \mathbf{t}')_K + (l, l')_K$$

$$= \int_{\mathbf{R}^3} [\mathbf{t}(k)/\mu(k)] \cdot \mathbf{t}'(k)^- \, dk +$$

$$+ m^2 \int_{\mathbf{R}^3} [\mathbf{l}(k)/\mu(k)] \cdot [\mathbf{l}'(k)/\mu(k)^2]^- \, dk.$$

Putting $u(-k) = \mathbf{t}'(k)^-$ and $v(-k) = m^2[\mathbf{l}'(k)/\mu(k)^2]^-$ and $\hat{f}(k) = u(k) + v(k)$ one sees easily, using (6.16), (6.8), and (6.10) that $\|f\|_{K^*} = \|h'\|_K$ and $\beta(\phi) = \langle f, \phi \rangle$.

THEOREM 6.2. *Denote by K_e the space of tempered distributions $g: \mathbf{R}^4 \to \mathbf{R}^4 \otimes \mathbf{C}$ such that*

$$\|g\|^2 = \|g\|^2_{-1} + m^{-2}\|\nabla \cdot g\|^2_{-1} < \infty. \tag{6.18}$$

For any Euclidean motion L on $\mathbf{R}^4$; $Lx = Ax + b$, where A is orthogonal and b is in $\mathbf{R}^4$, the induced action on K_e, given by $(L_*g)(x) = Ag(L^{-1}x)$, is unitary. Let K_e^0 denote the subspace of K_e consisting of those distributions with support in the hyperplane $x_4 = 0$. Then the map $f \to 2^{\frac{1}{2}} f \otimes \delta(x_4)$ is unitary from K^* onto K_e^0.

Put simply, the 'time-zero' subspace of K_e is K^*.

Proof. The norm (6.18) is clearly invariant under translations. If, moreover, $Lx = Ax$ is an orthogonal linear transformation, then the easily established equation $\nabla \cdot [Ag(A^{-1}x)] = (\nabla \cdot g)(A^{-1}x)$, along with the orthogonal invariance of the Sobolev norms shows that the norm (6.18) satisfies $\|L_*g\| = \|g\|$. Since $L_*(L^{-1})_*$ is the identity operator on K_e, L_* is unitary.

Now any complex valued distribution u in $\mathscr{S}'(\mathbf{R}^4)$ with support in the hyperplane $Y = \{(x, x_4) \in \mathbf{R}^4; x_4 = 0\}$ has the form $u = \sum_{j=0}^N v_j \otimes \delta^{(j)}(x_4)$, where v_j is in $\mathscr{S}'(Y)$ and N is the order of u. For example, v_j may be defined by $v_j(\phi) = u(\phi(x)x_4^j\psi(x_4))/j!$, where $\psi(x_4)$ is in $\mathscr{S}(\mathbf{R}^1)$ and is one in a neighbourhood of zero and ϕ is in $\mathscr{S}(Y)$. For if h is in $\mathscr{S}(\mathbf{R}^4)$ then $h - \sum_{j=0}^N (j!)^{-1}(\partial^j h(x,0)/\partial x_4^j)x_4^j\psi(x_4)$ along with its first N derivatives vanishes on Y and hence is annihilated by u, establishing the above representation. If in addition $\|u\|_{-1} < \infty$ then a computation shows $N = 0$ in the above representation; $u = v \otimes \delta(x_4)$. (We note that the preceeding discussion appears in [5; Proposition II.1].)

Suppose that g is in K_e^0. Then each component g_j is of the form $g_j = f_j \otimes \delta(x_4)$, where f_j is in $\mathscr{S}'(\mathbf{R}^3)$, since $\|g_j\|_{-1} < \infty$. But $\boldsymbol{\nabla} \cdot g = (\sum_{j=1}^{3} \partial f_j / \partial x_j) \otimes \delta(x_4) + f_4 \otimes \delta'(x_4)$. Since $\|\boldsymbol{\nabla} \cdot g\|_{-1} < \infty$ it follows from the preceding paragraph that $f_4 = 0$. Hence $g_4 = 0$. A straightforward computation, using $\int_{-\infty}^{\infty} (k_4^2 + a^2)^{-1} dk_4 = \pi a^{-1}$, now shows that $2\|f \otimes \delta\|_{K_e}^2 = \|f\|_{K^*}^2$, which concludes the proof.

Now, let J be the negative of time-reversal on K. Thus $(J\phi)_\alpha(t, x) = \phi^\alpha(-t, x)$. Then J is a conjugation on K. The J real substance of K, which by definition consists of those classical Proca fields ϕ such that $J\phi = \phi$, can be characterized as the set of those ϕ which at time zero have real spatial components and purely imaginary fourth component. We may identify K^* with K in a unitary manner by composing J with the natural anti-unitary map from a Hilbert space to its dual space. Note that, whereas the natural anti-unitary map from K to its abstract dual space is independent of Lorentz frame, the unitary map that we have just described between these spaces depends on the choice of a time-zero hyperplane in Minkowski space, because J is induced by the negative of reflection in this hyperplane. Of course, our explicit identification of the abstract dual space of K with a space of functions f in Theorem 6.1 also depended on the choice of a time zero hyperplane (on which the functions f live, in fact). One sees easily from (6.14) that the adjoint of J on K^* is simply complex conjugation; $f \to \bar{f}$. Upon identifying K with K^* in this unitary manner we may also extend this map to a unitary map from the symmetric Fock space $\mathscr{F}(K)$ over K to $\mathscr{F}(k^*)$. We henceforth identify $\mathscr{F}(K)$ and $\mathscr{F}(K^*)$ in this manner. Then the free Hamiltonian for the Proca field, which is an operator on $\mathscr{F}(K)$, may be identified with an operator H_0 on $\mathscr{F}(K^*)$. We write Re K^* for the real vector fields in K^* and similarly for K_e. Moreover, we identify K^* with the time-zero subspace of K_e via the unitary map described in Theorem 6.2.

Let θ: Re $K_e \to$ random variables be the normal distribution. That is, θ is the Gaussian process on real K_e with mean zero and covariance $E[\theta(f)\theta(g)] = (f, g)_{K_e}$, where E denotes expectation. If A is any closed subspace of Re K_e we write $L^2(A)$ for the space of square integrable functions on the sample space of θ which are measurable with respect to the σ-field generated by $\{\theta(g): g \in A\}$. As is well known [9] there is a unitary map from $\mathscr{F}(K^*)$ onto $L^2(\text{Re } K^*)$ which carries the time-zero quantized Proce field to multiplication operators on $L^2(\text{Re } K^*)$. We regard H_0 henceforth as an operator on $L^2(\text{Re } K^*)$.

THEOREM 6.3. *Let E_0 be the projection of $L^2(\text{Re } K_e)$ onto $L^2(\text{Re } K^*)$. Let T_s denote the induced (unitary) action in $L^2(\text{Re } K_e)$ of time translation in K_e;*

$T_s\theta(g) = \theta(g_s)$, where $g_s(x, x_4) = g(x, x_4 - s)$ and $x = (x_1, x_2, x_3)$. Then

$$\exp(-sH_0)u = E_0 T_s u, \qquad u \in L^2(\text{Re } K^*). \tag{6.19}$$

Moreover, let K_1 be the subspace of K_e consisting of those functions g in K_e whose fourth component is zero. Then the restriction of θ to K_1 is Markovian with respect to those half-spaces whose bounding hyperplane is of the form $\{x_4 = \text{constant}\}$.

Proof. Since $K^* \subset K_1$ and K_1 is closed in K_e and is also closed under time-translation (i.e. in the x_4 direction) as well as under complex conjugation the entire theorem is a theorem about K_1 rather than K_e. If f and g are in K_1 and are C^∞ functions with compact support then for each real s the function $f(x_1, x_2, x_3) \to f(x_1, x_2, x_3, s)$ in K^*. A computation shows that

$$(f, g)_{K_e} = \int_{-\infty}^{\infty} \int_{-\infty}^{\infty} 2^{-1}[e^{-|t-s|\mu} f(.,s), g(.,t)]_{K^*} \, dt \, ds. \tag{6.20}$$

Since K_1 is the closure of the space of such functions it follows that the time-translation group in K_1 is the minimal unitary extension of the semigroup $e^{-t\mu}$ in K^* ([5], [10], [15]). The theorem now follows just as in the case of a neutral scalar field ([5], [6], [10]).

REMARK 6.1. It appears that a physical interpretation for the real functions in K^* is that of instantaneous sources for the field. This comes out particularly clearly in the context of the electromagnetic field, to be discussed in the next section, where the elements in the dual space of the one particle space are identified with divergence-free currents on $\mathbf{R}^3$. The pairing $\langle f, \phi \rangle$ then has the interpretation of interaction energy of the current f with the field ϕ. Our norm in K^* agrees with the fundamental quadratic form for sources used by Schwinger for spin-one particles [11; eqns (2–3.4)].

REMARK 6.2. The subspace K_1 of K_e, defined in Theorem 6.3 is not invariant under four-dimensional Euclidean rotations. Thus the Euclidean theory is not directly tied to the one-particle source function space K^* via the theory of unitary dilations of contraction semigroups. However, a spin-s, $s = 1, 2, \ldots$, theory, with mass $m > 0$ has been constructed by Schrader and Uhlenbrock [10] based entirely on the theory of unitary dilations. Their Euclidean field in the spin-one case is a three-component field. It seems likely that their Euclidean field leads to a different notion of Euclidean locality from ours. Here it should be mentioned that T. H. Yao has recently constructed a Euclidean random process for spin one and mass $m > 0$ using analytic continuation methods. The author wishes to thank R. F. Streater for informing him of this.

Osterwalder and Schrader discuss Euclidean versions of higher spin theories [8]. They continue analytically to imaginary time the vacuum expectation values of products of the field ϕ and its first derivatives. A. H. Ozkaynak [16] has constructed free spin-s Euclidean fields in a Fock-space setting for $s = \frac{1}{2}, 1, \frac{3}{2}, \ldots$.

REMARK 6.3. It seems doubtful that the normal distribution θ on Re K_e is Markovian in the sense of Nelson [7]. The reason for this belief is that, by Theorem 6.2, one must expect that for any open set U in $\mathbf{R}^4$ with smooth boundary and any function f in K_e which is supported in ∂U, the component of f normal to ∂U will vanish at each point of ∂U. Thus the boundary Hilbert space suffers a loss of one degree of freedom and is too small to make the interior of U independent of the exterior. Specifically, Nelson's proof [6; Theorem 5] breaks down in the present setting in that the equation $[(m^2 - \Delta)g, h]_{K_e} = 0$ for all g in $C_c^\infty(\mathbf{R}^4; \mathbf{R}^4)$ with supt g in U'^0 does not imply $h = 0$ in U'^0. For example, if U'^0 is bounded and h is a smooth vector field with compact support such that $h_j(x_1, \ldots, x_4) = \exp(-mx_1)$ in U'^0 for $j = 1$ and $h_j = 0$ for $j = 2, 3, 4$, then $[(m^2 - \Delta)g, h]_{K_e} = 0$ for all of the above g. For a related discussion see [14; Appendix B].

6.3. The electromagnetic field

If one puts $m = 0$ in eqn (6.1) then the equations (6.1) and (6.2) are the equations for the electromagnetic field in the Lorentz gauge. The functions ϕ_α are of course the electromagnetic potentials. The construction of the one-particle Hilbert space proceeds in a manner similar to that for the Proca field, with a notable exception due to the degeneracy of the norm. Specifically, we define $\mathscr{C}_0 = \{k : k_0^2 = |\mathbf{k}|^2, k_0 > 0\}$ and T_k to be the tangent space to the light cone $\mathscr{C}_0$, at k, as before. But the Lorentz-invariant inner product is only positive-semidefinite on T_k. It has a one-dimensional singular subspace—namely, the longitudinal vectors, spanned by $k \in T_k$. Nevertheless we define M_0 to be the set of all measurable functions $h: \mathscr{C}_0 \to \mathbf{R}^4 \otimes \mathbf{C}$ such that

(a) $h(k) \in T_k \otimes \mathbf{C}$ for all k in $\mathscr{C}_0$,

(b) $\|h\|^2 = \int_{\mathscr{C}_0} h(k) \cdot \overline{h(k)} \frac{d^3k}{k_0} < \infty,$ (6.21)

and

(c) $\int_{\mathscr{C}_0} \sum_{j=0}^{3} |h_j(k)|^2 \frac{d^3k}{k_0} < \infty.$

The norm in (b) is degenerate in that $\|h\| = 0$ (not only if $h = 0$ almost everywhere but also if $h(k) = c(k)k$ for almost every k in $\mathscr{C}_0$, where $c(k)$ is a complex-valued function). Moreover these longitudinal functions constitute the kernel of $\| \ \|$. The one-particle Hilbert space M for Maxwell's equations may then be defined as the quotient space

$$M = M_0/(\text{kernel } \| \ \|).$$

Note that if $h(k) = t(k) + l(k)$ is the unique decomposition of $h(k)$ into a transverse and longitudinal part then $\|l\|^2 = 0$ and $\|h\|^2 = \|t\|^2 > 0$ unless $t(k) = 0$ a.e. It is clear that M can be identified with the purely transverse elements in M_0. Since the integral in (c) coincides with the integral in (b), when h is transverse, M is complete.

If h is in M_0, then, identifying $\mathscr{C}_0$ with its projection, $\mathbf{R}^3 - \{0\}$, in the hyperplane $\{k_0 = 0\}$, we see that

$$\int_{|k| \leq r} |h_j(k)|/|\mathbf{k}| \, d^3k \leq \left(\int_{|k| \leq r} |h_j(k)|^2/|\mathbf{k}| \, d^3k \right)^{\frac{1}{2}} \times$$

$$\times \left(\int_{|k| \leq r} |\mathbf{k}|^{-1} \, d^3k \right)^{\frac{1}{2}}$$

which is finite by (c). Thus (c) ensures that each of the components of $h(k)/|\mathbf{k}|$ is integrable on a neighbourhood of zero in $\mathbf{R}^3$ and square-integrable on its complement. Hence we may define

$$\phi(x) = (2\pi)^{-\frac{3}{2}} \int_{\mathscr{C}_0} h(k) \, e^{-ik \cdot x} \frac{d^3k}{k_0}, \quad x \in \mathbf{R}^4, \tag{6.22}$$

as a tempered distribution on $\mathbf{R}^4$. (In fact, for each t, $\phi(\mathbf{x}, t)$ is a sum of a bounded function and a square-integrable function on $\mathbf{R}^3$.) Of course ϕ satisfies eqns (6.1) and (6.2) with $m = 0$ in the distribution sense. The natural action of an orthochronous Lorentz transformation $Lx = Ax + b$ on ϕ is $(L_*\phi)(x) = A\phi(L^{-1}x)$, just as for the Proca field. This induces the action $(L_*h)(k) = e^{ik \cdot b} Ah(A^{-1}k)$ on M_0 as follows from (6.22) and the Lorentz-invariance of the measure d^3k/k_0 on $\mathscr{C}_0$. L_*h is in M_0 when h is in M_0 because (a) $AT_k = T_{Ak}$, (b) $\|L_*h\|^2 = \|h\|^2$, and

$$\text{(c)} \int |L_*h(k)|^2 \, d^3k/k_0 \leq \|A\|^2 \int |h(k)|^2 \, d^3k/k_0 < \infty,$$

where $|h(k)|^2 = \sum_{j=0}^{3} |h_j(k)|^2$ and $\|A\|$ denotes the norm of the Lorentz transformation A in the Euclidean norm of $\mathbf{R}^4$. L_* clearly takes the kernel of the norm (6.21)(b) into itself and therefore induces a unitary operator on M, yielding a unitary representation of the orthochronous Lorentz group. We note, however, that L_* does not necessarily take a purely transverse function h to another transverse function (i.e. the radiation gauge is not Lorentz-invariant). This is why the ambient space M_0 should be kept under consideration. To complete this expository account of the one-particle space for the Maxwell field we mention that time-reversal acts in M_0 and M via the anti-unitary operator induced by the map $\phi \to \phi'$, where $\phi'_\alpha(\mathbf{x}, t) = -\overline{\phi^\alpha}(\mathbf{x}, -t)$. We also remind the reader of the well-known fact [1] that the one-particle space can be described directly in terms of the electromagnetic field strengths E and H without the necessity of taking a quotient space. But the preceding description, in terms of the electromagnetic potentials, seems more suited for passing to the Euclidean setting.

The next three theorems are the respective analogues of Theorems 6.1, 6.2, and 6.3. The Sobolev norm of a vector-valued distribution on $\mathbf{R}^4$ is now defined as

$$\|f\|_a^2 = \int_{\mathbf{R}^n} |\hat{f}(k)|^2 |k|^{2a} \, dk, \qquad (6.23)$$

which is the same as (6.7) but with $m = 0$.

THEOREM 6.4. *Let M^* be the set of all distributions $f: \mathbf{R}^3 \to \mathbf{R}^3 \otimes \mathbf{C}$ such that*

$$\nabla \cdot f = 0 \qquad (6.24)$$

and

$$\|f\| = \|f\|_{-\frac{1}{2}} < \infty. \qquad (6.25)$$

Then M^* is dual to the one-particle electromagnetic field space M in the pairing

$$\langle f, \phi \rangle = \int_{\mathbf{R}^3} f(x) \cdot \phi(x, 0) \, dx \qquad (6.26)$$

Proof. By the Plancherel theorem

$$\langle f, \phi \rangle = \int_{\mathbf{R}^3} \hat{f}(-k) \cdot \mathbf{h}(k)/|k| \, dk, \qquad (6.27)$$

at least for a dense set of f and h. Since $k \cdot f(-k) = 0$ by (6.24), only the transverse part of h contributes to the right side of (6.27) so that the linear functional $\phi \to \langle f, \phi \rangle$ annihilates the kernel of the norm (6.21)(b). If h is transverse then $|\langle f, \phi \rangle| \leq (\int_{\mathbf{R}^3} |\hat{f}(-k)|^2 |k|^{-1} \, dk)^{\frac{1}{2}} \cdot (\int_{\mathbf{R}^3} |h(k)|^2 |k|^{-1} \, dk)^{\frac{1}{2}} = \|f\|_{M^*} \|h\|_M$. Hence the right side of (6.27) is a jointly continuous bilinear function on $M^* \times M$ whose continuous extension to all of $M^* \times M$ we take as the definition of the right side of (6.26). It is clear from (6.27) that M^* is the entire dual space of M.

THEOREM 6.5. *Let M_e be the Sobolev space of distributions $g: \mathbf{R}^4 \to \mathbf{R}^4 \otimes \mathbf{C}$ such that*

$$\nabla \cdot g = 0, \qquad (6.28)$$

$$\|g\| = \|g\|_{-1} < \infty. \qquad (6.29)$$

The natural action of the four-dimensional Euclidean group in M_e is unitary. The map $f \to 2^{\frac{1}{2}} f \otimes \delta(x_4)$ is unitary from K^* onto the subspace M_e^0 of distributions in M_e with support in the hyperplane $\{x_4 = 0\}$.

Proof. The Euclidean invariance of the norm (6.29) is clear and the invariance of the condition (6.28) follows again from the equation $\nabla \cdot [Ag(A^{-1}x)] = (\nabla \cdot g)(A^{-1}x)$. From the discussion in the proof of Theorem 6.2 we see that any distribution g with support in $\{x_4 = 0\}$ and with

$\|g\|_{-1} < \infty$ has the form $g = f \otimes \delta(x_4)$, where $f: \mathbf{R}^3 \to \mathbf{R}^4 \otimes \mathbf{C}$ in some tempered distribution. Then $\nabla \cdot g = (\nabla \cdot \mathbf{f}) \otimes \delta(x_4) + f_4 \otimes \delta'(x_4)$. Hence $\nabla \cdot g = 0$ implies $f_4 = 0$ and $\nabla \cdot \mathbf{f} = 0$. A computation shows $\|g\|_{-1} = 2^{\frac{1}{2}} \|\mathbf{f}\|_{-\frac{1}{2}}$. Therefore the map $f \to 2^{\frac{1}{2}} f \otimes \delta(x_4)$ is isometric from K^* onto M_e^0.

THEOREM 6.6. *All of the conclusions of Theorem 6.3 hold when K_e is replaced by M_e, K^* by M^*, and K_1 by the set M_1 of all divergence free vector fields in M_e with vanishing fourth component.*

Proof. The proof is the same as that of Theorem 6.3 with the exception that the functions f, g for which one should establish eqn (6.20) are those divergence-free vector fields f in M_1 such that $\hat{f}(k) = 0$ if $|\mathbf{k}| < \varepsilon$ or $|k| > \varepsilon^{-1}$. These f are dense in M_1 while $f(.,s)$ is in M^* for each real s.

REMARK 6.4. The elements of M^* can be interpreted as current densities, i.e. 'test currents' with charge density zero. The condition $\nabla \cdot f = 0$ is commensurate with constant charge density zero. For example, a test body for the free electromagnetic field consisting of a closed loop of wire carrying a current can be represented at time zero by an element of M^*. The right side of (6.26) is then the interaction energy of the loop with the field ϕ at time zero.

REMARK 6.5. The problem of gauge-invariance manifests itself for the free electromagnetic field in that the elements of M are not in one-to-one correspondence with solutions ϕ of (6.1)-(6.2) (because of the degeneracy of the M_0 norm) while the transverse elements of M_0 (which are in one-to-one correspondence with M) do not form an invariant set in M_0 under Lorentz transformations. It seems remarkable therefore that the Euclidean test-function space M_e, which has a Euclidean-invariant and non-degenerate norm, does not reflect the gauge-invariance problems of the relativistic setting, in spite of the fact that the construction of M_e is based on the use of the electromagnetic potentials rather than the field strengths. It remains to be seen whether this will persist when interactions are considered.

REMARK 6.6. Similar results hold for higher-integer-spin fields. The Euclidean test-functions for spin $s \geq 2$ must satisfy second-order differential equations, however. These results will be presented elsewhere.

REMARK 6.7. The original construction of the Euclidean electromagnetic field appears in Schwinger's papers [12, 13].

REMARK 6.8. The comments in Remark 6.3 concerning the failure of the Markoff property apply to the electromagnetic field as well. An example for the breakdown of Nelson's proof [6; Theorem 5] in the present setting is supplied by the function $h = \nabla u$, where u is a C^∞ real function on $\mathbf{R}^4$ with compact support and which is harmonic in the bounded open set U'^0. For such a function one has $(-\Delta g, h)_{M_e} = 0$ for all C^∞ functions g in M_e with compact support in U'^0. h is in M_e because $\nabla \cdot h = \nabla \cdot \nabla u = 0$.

6.4. Note: (added November, 1974)

T. H. Yao has shown (Bedford College preprint, London, July 1974) that the Euclidean Proca field (spin 1, $m > 0$) is indeed Markovian, contrary to the implied conjecture in Remark 6.3 of this paper. His clever proof consists in showing, with the use of Fourier transforms, that there exists a local differential operator Q such that $(Qu, v)_{K_e} = (u, v)_{L^2}$ for suitable u and v in K_e, so that Nelson's method is now applicable. The operator Q may, in fact, be described explicitly as follows.

Regard a vector field on R^4 to be a one-form. Then we may use the exterior derivative operator d and its adjoint δ (in $L^2(R^4, \Lambda(R^4))$) to describe operators succinctly in the following well known way: $-\Delta = d\delta + \delta d$ and divergence $u = \delta u$ if u is a one-form. By eqn (2.18)

$$(u, v)_{K_e} = (u, v)_{-1} + m^{-2}(\delta u, \delta v)_{-1} = ((1 + m^{-2} d\delta)u, v)_{-1}$$

for reasonable u and v in K_e. Put $Q = m^2 + \delta d$, restricted to one forms. Then, since $d^2 = \delta^2 = 0$ we have $(1 + m^{-2} d\delta)Q = m^2 - \Delta$. Hence

$$(Qu, v)_{K_e} = ((m^2 - \Delta)u, v)_{-1} = (u, v)_{L^2}.$$

The Markov property for the Euclidean electromagnetic field is still an open question.

Acknowledgement

This work has been supported in part by N.S.F. Grant GP31380.

References

1. BARGMANN, V. and WIGNER, E. P. Group theoretical discussion of relativistic wave equations. *Proc. natn. Acad. Sci. U.S.A.* **34**, 211–23 (1948).
2. FRÖHLICH, J. Schwinger functions and their generating functionals, I and II. Harvard preprint (1973).
3. FRÖHLICH, J. Verification of axioms for Euclidean and relativistic fields and Haag's theorem in a class of $P(\phi)_2$ models. Harvard preprint (1974).
4. GLIMM, J., JAFFE, A., and SPENCER, T. The particle structure of the weakly coupled $P(\phi)_2$ model and other applications of high temperature expansions, Part I: Physics of Quantum field models. In *Constructive quantum field models*, (eds G. Velo and A. Wightman). Springer–Verlag, Berlin (1973).
5. GUERRA, F., ROSEN, L., and SIMON, B. The $P(\phi)_2$ Euclidean quantum field theory as classical statistical mechanics. *Ann. Math.* (To appear.)
6. NELSON, E. The free Markoff field. *J. funct. Anal.* **12**, 211–27 (1973).
7. NELSON, E. Construction of quantum fields from Markoff fields. *J. funct. Anal.* **12**, 97–112 (1973).

8. OSTERWALDER, K. and SCHRADER, R. Axioms for Euclidean Greens functions. *Commun. Math. Phys.* **31**, 83–112 (1973).
9. SEGAL, I. E. Tensor algebras over Hilbert spaces. *Trans. Am. math. Soc.* **81**, 106–34 (1956).
10. SCHRADER, R. and UHLENBROCK, D. A. Markov structures on Clifford algebras. Freie Universität Berlin preprint (1974).
11. SCHWINGER, J. *Particles, sources and fields.* Addison–Wesley, Reading, Mass. (1970).
12. SCHWINGER, J. On the Euclidean structure of relativistic field theory. *Proc. natn. Acad. Sci. U.S.A.* **44**, 956–65 (1958).
13. SCHWINGER, J. Euclidean quantum electrodynamics. *Phys. Rev.* **115**, 721–31 (1959).
14. SYMANZIK, K. W. A modified model of Euclidean quantum field theory. New York University preprint (1964).
15. SZ.-NAGY, B. and FOIAS, C. *Harmonic analysis of operators on Hilbert space.* North-Holland, London (1970).
16. OZKAYNAK, A. H. Ph.D. Thesis, University of Harvard (1974). (Private communication.)

7. The design of future computing machinery for functional integration

J. M. HAMMERSLEY

THE need of any physicist, or more generally any scientist, is to *understand* the implications of his theories. These implications are often quantitative, and hence expressible in terms of numbers produced by a computer. However, numbers are *not* readily understandable, particularly if they occur in bulk. When a computer has churned out, say, 50 pages of typewritten numbers, each to 6 significant figures, it can be very difficult to see the wood for the trees. Numbers can only be fed into the mind one at a time, and assembled there rather laboriously. Thus the problem concerns the interface between the machine and the human mind.

The concept of a computer as simply a number-cruncher and a printer of tabulations, however established and current as a concept, is also an outdated idea, because we already possess the technology to do better. The main machinery of the human mind, at least in the realm of intellectual thought, consists of our organs of sight and sound. We find it very much easier to grasp a situation in its synoptic whole if we can see it diagrammatically or if we can hear it thematically. These human facilities are often the fruit of experience or training, sometimes acquired unconsciously through the mere business of living in a complex world that requires quick responses, and sometimes acquired through deliberate and laborious learning. They result from combining the brute sensations of sight and sound with the interpretative powers of the brain. For example, we all possess the commonplace but fabulous means to interpret the three-dimensional world around us from a two-dimensional image on the retina—you can do it with one eye shut. We can pick out significant noises from the midst of a wildly complicated pattern of sounds: we are instantly aware when the car engine knocks, or a *Leitmotiv* occurs in the middle of the racket of *Götterdammerung*. A trained cartographer can absorb as much information from a 5-second glance at an ordnance survey map as would occupy him to describe in a half-hour telephone conversation.

A computer can pour out information (in the information–theoretic sense) very rapidly; but this speed is not properly utilized unless our minds can take it in equally rapidly. So we need to attune the computer output to our rapid human input mechanisms, eye and ear; and we need to do it in the same sense as we might tune or match some electronic transmission line. Work along these lines is already in progress: it involves collaboration between electronic engineers, computer designers, and experimental psychologists. They are attaching television output to the computer and investigating how best to use colour and stereoscopy, coupled perhaps with loudspeaker sound, so that the selective attention of the human observer is most appropriately caught. Should you catch it by movement or changes in colour, or intensity, or perspective, or pitch, or what else?

Some applied mathematicians are already making use of visual outputs from computers. The lead seems to be coming, in particular, from magnetohydrodynamicists (they have good financial resources for the most part). They have also been active in promoting work on the other main computer development which promises well for the future, namely the advent of the multiminicomputer. Early parallel computers, like Illiac IV, have been somewhat limited by the electronic technology available when they were designed some years ago. But large-scale integrated circuitry is going to revolutionize the field. Suppose we wish to solve a set of partial differential equations, say of evolutionary type with one time variable and two space variables. We can set up a two-dimensional grid for the space variables, and we can compute the solution at each grid point at successive instants of time. An old-fashioned serial machine would store the grid in its memory, would bring values of the function at these grid points to its single central processor one at a time for updating by recalculation, and would then return them to the memory. The idea of the multiminicomputer, on the other hand, is to have a separate minicomputer at each grid point in the memory; and to update all grid points simultaneously in parallel. A typical multiminicomputer, at present under study by ICL, has 18 432 minicomputers at the grid points of a 128×144 array; and each of these minicomputers has its own little processor with a 40 ns multiplication time and with 4096 bits of internal storage. That is an encouraging start; but, of course, it does not go as far as we would wish. We really need a three-dimensional array, because the interesting and important physical problems are three-dimensional. Suppose we think in terms of an array of size $128 \times 128 \times 128$, giving 2 097 152 grid points in all. The current price for a minicomputer of the type considered is about 20 dollars; so a multiminicomputer with about 2 million grid points would cost 40 million dollars, which is expensive though not impossibly so. Moreover, with development the cost of each minicomputer is likely to fall, perhaps even by a factor of 10 when we start mass-producing them in millions, and we can probably look forward to machines of this sort of size in the fairly near future.

From the description of the machine so far, you will visualize a three-dimensional rectangular array of about 2 million minicomputers, each minicomputer connected to the 6 neighbouring minicomputers and able to transmit data to or from its 6 neighbours. These interconnections impose a fixed cubical topology on the machine; and a fixed topology is undesirable. For instance, we might want to study the probabilistic structure of cooperative phenomena on a crystal lattice. That would be admirable if the crystal were simple cubic; but would land us in trouble with (say) a tetrahedral lattice. If we were handling a still more complicated problem, such as melting ice, we should want a tetrahedral topology inside the ice together with a fluid topology in the water. In the liquid phase, we might want each minicomputer to travel around connecting itself up to other liquid elements as though pursuing the Lagrangian formulation of the motion. We can achieve this flexible topology by providing each minicomputer with an adequate battery of index registers, say 16 or 32 index registers for each; and these index registers can keep a continuous tally of which other minicomputers are currently its neighbours. It goes without saying that the requisite electronics are more complex than this simple account would appear to suggest; but, doubtless, the engineers will cope in due course. One needs at least 12 index registers to cope with the face-centred cubic lattice. Moreover, since the average number of faces of a random Veronoi tesselation is 15·54, even 16 index registers might well prove inadequate for problems in stochastic geometry.

Do these developments have relevance for functional integration? On the surface, it seems so. For instance, the grid points of the multiminicomputer can store functional values; we can trace paths across the grid by connecting up grid points via index registers; and hence we can sum these values to give path-integrals. Moreover, we could sum over paths, varying the paths through the flexibility of the indexably associative topology. Again, a number of functional-analytic problems are essentially exercises in the calculus of variations; i.e. path-variations of the kind just mentioned.

Ultimately these issues depend upon what problems the functional analyst wants to solve, and to what extent numerical or diagrammatic representations will help the physicist to understand his physical theories. These are questions which we must leave to others to answer. However, if there are opportunities here, and especially if there are peculiar functional-analytic requirements which might call for particular modifications in computer design, now is the time to make these requirements known to computer designers while they are still in the early stages of their designing and thus more easily able to take them into account.

For we must remember that the designers of a multiminicomputer capable of parallel indexably-associative network operation (PIANO) have a very considerable technological problem on their hands. Such a computer resembles a telephone network, and we may visualize its complexity by comparing it with the telephone system in England, say. Here we have a few million subscribers, any one of whom may communicate (through STD dialling) with any other. For this end, there are various switching devices in local exchanges; and these switches are so designed to take account of the fact that most subscribers use the telephone only infrequently, and even then most usually to make only local calls. For PIANO we have to visualize all subscribers in continuous conversation, each with say 32 other subscribers;

and, while these conversations are instantaneously local, the localities are in continuous interchangeable and intermiscible motion. It is as though bits of Bootle were drifting apart, some bits towards Plymouth, other bits to Scarborough; and meanwhile bits of Plymouth are headed towards Margate; and so on and so forth. And all the switching has to be ultra-high-speed; for we are computing at nanosecond rates. These switching problems are going to cause some real technological headaches; but equally the author does not doubt that they will be solved eventually because the rate of electronic advances shows little sign of any faltering. But *just* how they will be solved will depend a good deal on the type of *user demand* there is from various sorts of applied mathematicians and scientists. Do functional analysts need to get in on the act?

A fuller account of the developments described here appears in reference [1].

Reference

1. HAMMERSLEY, J. M. Some speculations on a sense of nicely calculated chances. *Rev. Soc. ind. appl. Math.* **16**. (In the press.)

8. Some probabilistic aspects of scattering theory

M. KAC AND P. VAN MOERBEKE

8.1. Introduction

IN this chapter we shall discuss a number of connections between integration in function spaces and problems of scattering theory. We shall restrict ourselves to the one-dimensional case, and we shall not strive for ultimate generality. Our purpose is mainly to point out the various advantages of the probabilistic view rather than to try and obtain the best results. Scattering theory is concerned with the following problem.

Consider the Schrödinger equation

$$\phi'' - q(x)\phi = -\lambda\phi(x), \tag{8.1}$$

where we shall assume that the potential $q(x)$ is sufficiently smooth and vanishes outside the interval $(-a, a)$ (this is not necessary, but it greatly simplifies the discussion).

For $\lambda = k^2 > 0$, one seeks the solution $f(x; \lambda)$ of (8.1) which has the following asymptotic behaviours at infinity,

$$f(x; \sqrt{\lambda}) \sim e^{-i\sqrt{(\lambda)}x} + b(\sqrt{\lambda}) e^{i\sqrt{(\lambda)}x}, \qquad x \to \infty \tag{8.2a}$$

and

$$f(x; \sqrt{\lambda}) \sim a(\sqrt{\lambda}) e^{-i\sqrt{(\lambda)}x}, \qquad x \to -\infty \tag{8.2b}$$

(in our case of q vanishing outside an interval the asymptotic relations above become strict equalities for sufficiently large $|x|$).

The coefficients $a(\sqrt{\lambda})$ and $b(\sqrt{\lambda})$ are called the transmission and reflection coefficients respectively, and they satisfy the relationship

$$|a(\sqrt{\lambda})|^2 + |b(\sqrt{\lambda})|^2 = 1. \tag{8.3}$$

It is convenient to introduce functions $f_1(x; k) = f_1(x; \sqrt{\lambda})$ and $f_2(x; k) = f_2(x; \sqrt{\lambda})$ which are defined (uniquely) by the asymptotic relations

$$f_1(x; k) \sim e^{ikx}, \qquad x \to \infty \tag{8.4a}$$

and

$$f_2(x; k) \sim e^{-ikx}, \qquad x \to -\infty \tag{8.4b}$$

and to accept the convention

$$f_1(x; -k) = f_1^*(x; k) \tag{8.5a}$$
$$f_2(x; -k) = f_2^*(x; k). \tag{8.5b}$$

To further simplify the discussion, we shall assume that there are no bound states, i.e. no square-integrable solutions of (8.1) corresponding to negative λ.

The fundamental solution $P(x|y; t)$ of the equation

$$\frac{\partial P}{\partial t} = \frac{\partial^2 P}{\partial x^2} - q(x)P \qquad (8.6)$$

can now be written in the form

$$P(x|y; t) = \frac{1}{2\pi}\int_{-\infty}^{\infty} \exp(-k^2 t)[f_1(x; k)f_1(y; -k) + f_1(x; k)f_1(y; k)f(k)]\, dk$$

$$= \frac{1}{2\pi}\int_{-\infty}^{\infty} \exp(-k^2 t)a(k)f_1(y; k)f_2(x; k)\, dk \qquad (8.7)$$

while, on the other hand,

$$P(x|y; t) = \frac{\exp[-(y-x)^2/4t]}{2\sqrt{(\pi t)}} E\left(\exp\left\{-\int_0^t q[y + x(\tau)]\, d\tau\right\}\bigg| x(t) = y - x\right) \qquad (8.8)$$

where $x(\tau)$ is Brownian motion normalized in a slightly different way to account for the coefficient 1 of ϕ'' in (8.1) rather than for the more customary $\frac{1}{2}$.

It is the interplay between the classical formula (8.7) and the probabilistic formula (8.8) that is at the heart of our discussion.

8.2. A trace formula

As an illustration of the interplay mentioned above, we shall sketch a derivation of the simplest trace formula (see e.g. [1]). All such trace formulas can be derived by this method, but one will suffice for our purposes.

If we take $x > a$ and $y < -a$, we find by combining formulas (8.7) and (8.8) that

$$E\left(\exp\left\{-\int_0^t q[x + x(\tau)]\, d\tau\right\}\bigg| x + x(t) = y\right)$$

$$= \frac{\sqrt{t}}{\sqrt{\pi}}\int_{-\infty}^{\infty} \exp\left[-\left(k\sqrt{t} + i\frac{y-x}{2\sqrt{t}}\right)^2\right] a(k)\, dk$$

$$= \frac{1}{\sqrt{\pi}}\int_{-\infty}^{\infty} \exp\left[-\left(k + i\frac{y-x}{2\sqrt{t}}\right)^2\right] a\left(\frac{k}{\sqrt{t}}\right) dk. \qquad (8.9)$$

The idea is now to expand both sides in powers of t (for small t) and compare the coefficients.

To lowest order in t the E term easily yields

$$1 + \frac{t}{(x-y)} \int q(u)\,du, \tag{8.10}$$

but the integral involving the transmission coefficient $a(k)$ requires a number of (non-trivial) analytic facts.

First we need the fact that $a(k)$ is analytic in the upper k half-plane and that, moreover, it is related to the reflection coefficient by the formula

$$a(k) = \exp\left[\frac{1}{2\pi i}\int_{-\infty}^{\infty}\frac{\ln(1-|b(u)|^2)}{u-k}\,du\right]. \tag{8.11}$$

Next, proceeding formally, we transform the right-hand side integral of (8.9) into

$$\frac{1}{\sqrt{\pi}}\int_{-\infty}^{\infty}\exp(-x^2)a\left(\frac{k}{\sqrt{t}} - i\frac{y-x}{2t}\right)dk,$$

and using (8.11) we obtain that for small t it is asymptotically equal to

$$1 + \frac{1}{\pi}\frac{t}{y-x}\int_{-\infty}^{\infty}\ln(1-|b(u)|^2)\,du. \tag{8.12}$$

This upon comparing with (8.10) yields

$$\int_{-\infty}^{\infty} q(u)\,du = -\frac{1}{\pi}\int_{-\infty}^{\infty}\ln(1-|b(u)|^2)\,du \tag{8.13}$$

which is the simplest trace formula.

8.3. Introducing the Konteweg–de Vries flow

Let us now turn to a problem which has only an indirect bearing on scattering theory but is of considerable independent interest.

We have in mind a type of inverse Sturm–Liouville problem and its connection with the Konteweg–de Vries equation.

Let $q(\alpha; x)$ be a family of smooth periodic (in x) potentials of period a,

$$q(\alpha; x+a) = q(\alpha; x), \tag{8.14}$$

and consider the eigenvalue problem

$$\frac{1}{2}\frac{d^2\phi}{dx^2} - q(\alpha; x)\phi = -\lambda(\alpha)\phi \tag{8.15}$$

with periodic boundary conditions.

It is known [2], and the authors consider it to be one of the most interesting discoveries of recent years, that if the potential $q(\alpha; x)$ evolves according to the Konteweg–de Vries equation

$$\frac{\partial q}{\partial \alpha} = \frac{\partial^3 q}{\partial x^3} - 12q\frac{\partial q}{\partial x} \qquad (8.16)$$

the eigenvalues corresponding to our problem remain unchanged!

Let us give a partly probabilistic, partly analytic proof of this remarkable fact.

Let $\tilde{x}(t)$ be the Wiener process mod a, i.e.

$$\tilde{x}(t) = x(t) - a\left[\frac{x(t)}{a}\right], \qquad (8.17)$$

where $x(t)$ ($x(0) = 0$) is the standard Wiener process.

We have

$$\sum_{n=1}^{\infty} \exp[-\lambda_n(\alpha)t] = \int_0^a E\left(\exp\left\{-\int_0^t q[\alpha; x + \tilde{x}(u)]\,du\right\}; x + \tilde{x}(t)\varepsilon\,dx\right), \qquad (8.18)$$

with the understanding that $x + \tilde{x}(t)\varepsilon\,dx$ means that $x + \tilde{x}(t)$ is in the differential (dx) interval about x. Differentiating (8.18) with respect to α we obtain

$$\frac{d}{d\alpha} \sum_{n=1}^{\infty} \exp[-\lambda_n(\alpha)t]$$

$$= -\int_0^a E\left(\int_0^t \frac{\partial}{\partial \alpha} q[\alpha; x + \tilde{x}(\tau)]\,d\tau \exp\left\{-\int_0^t q[\alpha; x + \tilde{x}(u)]\,du\right\}; x + \tilde{x}(t)\varepsilon\,dx\right)$$

$$= -\int_0^t d\tau \int_0^a E\left(\frac{\partial}{\partial \alpha} q[\alpha; x + \tilde{x}(\tau)]\exp\left\{-\int_0^t q[\alpha; x + \tilde{x}(u)]\,du\right\}; x + \tilde{x}(t)\varepsilon\,dx\right)$$

$$= -\int_0^t d\tau \int_0^a E\left(\frac{\partial}{\partial \alpha} q[\alpha; x + \tilde{x}(\tau)]\exp\left\{-\int_0^\tau q[\alpha; x + \tilde{x}(u)]\,du\right\} \times \right.$$
$$\left. \times \exp\left\{-\int_\tau^t q[\alpha; x + \tilde{x}(u)]\,du\right\}; x + \tilde{x}(t)\varepsilon\,dx\right). \qquad (8.19)$$

Now, classifying the paths $x + \tilde{x}(u)$, $0 \leq u \leq t$, according to which differential (dy) window they pass through at time τ and using the Markovian property

Some probabilistic aspects of scattering theory   91

(as well as temporal homogeneity) of our process, we have

$$E\left(\frac{\partial}{\partial\alpha}q[\alpha;x+\tilde{x}(\tau)]\exp\left\{-\int_0^\tau q[\alpha;x+\tilde{x}(u)]\,du\right\}\times\right.$$
$$\left.\times\exp\left\{-\int_\tau^t q[\alpha;x+\tilde{x}(u)]\,du\right\};x+\tilde{x}(t)\varepsilon\,dx\right)$$
$$=\int_0^a\frac{\partial}{\partial\alpha}q(\alpha;y)E\left(\exp\left\{-\int_0^\tau q[\alpha;x+\tilde{x}(u)]\,du\right\};x+\tilde{x}(\tau)\varepsilon\,dy\right)$$
$$=E\left(\exp\left\{-\int_0^{t-\tau}q[\alpha;y+\tilde{x}(u)]\,du\right\};y+\tilde{x}(t-\tau)\varepsilon\,dx\right), \qquad (8.20)$$

the integration being clearly with respect to y.

We also have

$$E\left(\exp\left\{-\int_0^\tau q[\alpha;x+\tilde{x}(u)]\,du\right\};x+\tilde{x}(\tau)\varepsilon\,dy\right)$$
$$=\sum_{n=1}^\infty\exp[-\lambda_n(\alpha)\tau]\phi_n(\alpha;x)\phi_n(\alpha;y)\,dy \qquad (8.21a)$$

and

$$E\left(\exp\left\{-\int_0^{t-\tau}q[\alpha;y+\tilde{x}(u)]\,du\right\};y+\tilde{x}(t-\tau)\varepsilon\,dx\right)$$
$$=\sum_{n=1}^\infty\exp[-\lambda_n(\alpha)(t-\tau)]\phi_n(\alpha;y)\phi_n(\alpha;x)\,dx, \qquad (8.21b)$$

which when substituted in (8.20) and then in (8.19) finally lead to the basic formula

$$\frac{d}{d\alpha}\sum_{n=1}^\infty\exp[-\lambda_n(\alpha)t]=-t\sum_{n=1}^\infty\exp[-\lambda_n(\alpha)t]\int_0^a\frac{\partial}{\partial\alpha}q(\alpha;y)\phi_n^2(\alpha;y)\,dy, \qquad (8.22)$$

where it is, of course, understood that the ϕ_n are the normalized eigenfunctions of (8.15) (with periodic boundary conditions).

The above derivation is essentially that of Melnikoff [3].

8.4. Konteweg–de Vries flow continued

In this section the discussion is elementary, purely analytic, and is directed toward proving that, for all n,

$$\int_0^a\left(\frac{d^3q}{dx^3}-12q\frac{dq}{dx}\right)\phi_n^2(x)\,dx=0 \qquad (8.23)$$

(since the dependence on α is irrelevant for the purpose of this discussion it is suppressed). Multiplying both sides of (8.15) by ϕ' and integrating from 0 to a with respect to x we obtain (using periodic boundary conditions)

$$\int_0^a q(x)\frac{d}{dx}\phi_n^2\,dx = 0,$$

or, integrating by parts (and again using periodic boundary conditions as well as periodicity of q),

$$\int_0^a \frac{dq}{dx}\phi_n^2(x)\,dx = 0. \tag{8.24}$$

Next we differentiate (8.10) (with ϕ replaced by ϕ_n) with respect to x obtaining

$$\frac{1}{2}\phi_n''' - q(x)\phi_n' - \frac{dq}{dx}\phi_n = -\lambda_n\phi_n', \tag{8.25}$$

and multiply (8.25) by ϕ_n'' and integrate on x from 0 to a. We obtain

$$-\frac{1}{2}\int_0^a q(x)\frac{d}{dx}(\phi_n'^2)\,dx - \int_0^a \frac{dq}{dx}\phi_n\phi_n''\,dx = 0$$

or

$$\frac{1}{2}\int_0^a \frac{dq}{dx}\phi_n'^2(x)\,dx = 2\int_0^a \frac{dq}{dx}\phi_n[q(x)\phi_n - \lambda_n\phi_n]\,dx = 2\int_0^a q\frac{dq}{dx}\phi_n^2\,dx, \tag{8.26}$$

where use has been made of (8.24).

On the other hand (again using (8.24)),

$$0 = -\lambda_n \int_0^a \frac{dq}{dx}\phi_n^2(x)\,dx = \int_0^a \frac{dq}{dx}\phi_n(x)[\tfrac{1}{2}\phi_n'' - q(x)\phi_n]\,dx$$

$$= -\int_0^a q\frac{dq}{dx}\phi_n^2\,dx + \frac{1}{2}\int_0^a \frac{dq}{dx}\phi_n\phi_n''\,dx$$

$$= -\int_0^a q\frac{dq}{dx}\phi_n^2\,dx - \frac{1}{2}\int_0^a \left[\frac{d^2q}{dx^2}\phi_n + \frac{dq}{dx}\phi_n'\right]\phi_n'\,dx$$

$$= -\int_0^a q\frac{dq}{dx}\phi_n^2\,dx - \frac{1}{4}\int_0^a \frac{d^2q}{dx^2}\frac{d}{dx}(\phi_n^2)\,dx - \frac{1}{2}\int_0^a \frac{dq}{dx}\phi_n'^2\,dx$$

$$= \int_0^a \left(\frac{1}{4}\frac{d^3q}{dx^3} - q\frac{dq}{dx}\right)\phi_n^2\,dx - \frac{1}{2}\int_0^a \frac{dq}{dx}\phi_n'^2\,dx$$

$$= \int_0^a \left(\frac{1}{4}\frac{d^3q}{dx^3} - q\frac{dq}{dx}\right)\phi_n^2\,dx - 2\int_0^a \frac{dq}{dx}\phi_n^2\,dx = \frac{1}{4}\int_0^a \left(\frac{d^3q}{dx^3} - 12q\frac{dq}{dx}\right)\phi_n^2\,dx,$$

where use has been made of (8.26).

Thus (8.24) follows, and hence if

$$\frac{\partial q}{\partial \alpha} = \frac{\partial^3 q}{\partial x^3} - 12q\frac{\partial q}{\partial x}, \qquad (8.27)$$

we have by (8.22),

$$\frac{d}{d\alpha} \sum_{n=1}^{\infty} \exp[-\lambda_n(\alpha)t] = 0 \qquad (8.28)$$

for *all* $t>0$, which implies that the individual eigenvalues do not change under the flow (8.27).

8.5. A question of isospectrality

In general the eigenvalues of (8.15) (corresponding to periodic boundary conditions) do not determine the potential q.

There are, however, special potentials which are uniquely determined (up to a translation, of course) by their spectra.

Starting with the formula (8.18) rewritten in terms of a conditional expectation)

$$\sum_{n=1}^{\infty} \exp(-\lambda_n t) = \int_0^a P(x|x;t) E\left(\exp\left\{-\int_0^t q[x+\tilde{x}(u)]\,du\right\}\bigg|\tilde{x}(t)=0\right) dx, \qquad (8.29)$$

where

$$P(x|y;t) = \frac{1}{\sqrt{(2\pi t)}} \sum_{m=-\infty}^{\infty} \exp\left[-\frac{(y-x-ma)^2}{2t}\right], \qquad (8.30)$$

we derive the asymptotic formula for (8.29) for small t.

This is done most conveniently by writing

$$\int_0^t q[x+\tilde{x}(u)]\,du = tq(x) + \frac{q'(x)}{1!}\int_0^t \tilde{x}(u)\,du + \ldots, \qquad (8.31)$$

and then expanding the exponential.

Proceeding in this way, we obtain

$$\sum_{n=1}^{\infty} \exp(-\lambda_n t) \sim \frac{a}{\sqrt{(2\pi t)}}\left\{1 - \frac{t}{a}\int_0^a q(x)\,dx + \frac{t^2}{2}\frac{1}{a}\int_0^a q^2(x)\,dx - \frac{t^3}{24}\frac{1}{a}\int_0^a [q'^2(x) + 4q^3(x)]\,dx + \ldots\right\}, \qquad (8.32)$$

where use has been made of asymptotic formulas

$$E\left\{\left[\int_0^t \tilde{x}(u)\,du\right]^2 \bigg| \tilde{x}(u)=0\right\} \sim \frac{t^3}{12},$$

$$E\left\{\int_0^t \tilde{x}^2(u)\,du \bigg| \tilde{x}(u)=0\right\} \sim \frac{t^2}{6}$$

(for more details, see [4]).

In a suitably chosen space of potentials, it can be shown that there is a unique (up to a translation) q which minimizes

$$\int_0^a [q'^2(x)+4q^3(x)]\,dx$$

under the constraints

$$\frac{1}{a}\int_0^a q(x)\,dx = A, \qquad \frac{1}{a}\int_0^a q^2(x)\,dx = B.$$

Such a potential (which is easily seen to be a Legendre elliptic function) is uniquely determined (up to a translation) by its spectrum because of the unicity of the solution of the variational problem.

8.6. A discrete analogue

Consider a random walk with equiprobable ± 1 steps on vertices of a regular N-gon and let $q_t(i)$, $1 \le i \le N$, be a one parameter family of functions defined on it (we use t instead of α to emphasize a connection with classical mechanics which will emerge a little later).

Consider the matrix $Q(t)$ whose elements are given by the formula

$$Q_{xy}(t) = \exp[-\tfrac{1}{2}q_t(x)]P(x|y)\exp[-\tfrac{1}{2}q_t(y)], \qquad (8.33)$$

where x, y are integers from 1 to N and

$$P(x|y) = \begin{cases} \tfrac{1}{2}, & \text{if } y=x+1 \text{ or } x-1 \\ 0, & \text{otherwise} \end{cases} \qquad (8.34)$$

(it is, of course, understood that $N+1$ is to be read as 1 and -1 as $N-1$).

Denoting by $\lambda_j(t)$, $j=1, 2, \ldots, N$, the eigenvalues of $Q(t)$, we have

$$\sum_{j=1}^N \lambda_j^n(t) = \sum_{s_0=1}^N E\left\{\exp\left[-\sum_{k=0}^{n-1} q_t(s_k)\right]; s_n = s_0\right\}, \qquad (8.35)$$

which is the analogue of (8.18).

Differentiating (8.35) with respect to t, we obtain by imitating the argument of §8.3 the formula

$$\frac{d}{dt}\left[\sum_{j=1}^{N} \lambda_j^n(t)\right] = -n \sum_{j=1}^{N} \lambda_j^n(t) \sum_{y=1}^{N} \phi_j^2(y) \frac{dq_t(y)}{dt}, \quad (8.36)$$

where $\phi_j(y)$, $y = 1, 2, \ldots, N$ is the eigenvector of Q belonging to $\lambda_j(t)$ (its dependence on t has been suppressed).

Formula (8.36) is a clear analogue of (8.22).

Suppressing subscripts and dependence on t, we write the eigenvalue equations in the explicit form

$$\tfrac{1}{2}\exp\left[-\frac{q(x-1)+q(x')}{2}\right]\phi(x-1) + \tfrac{1}{2}\exp\left[-\frac{q(x)+q(x+1)}{2}\right]\phi(x+1)$$
$$= \lambda \phi(x)$$

or setting

$$\exp\left[-\frac{q(x)}{2}\right]\phi(x) = \psi(x) \quad (8.37)$$

in the equivalent form

$$\tfrac{1}{2}\psi(x-1) + \tfrac{1}{2}\psi(x+1) = \lambda e^{q(x)}\psi(x). \quad (8.38)$$

Multiplying both sides of (8.38) by $e^{-q(x)}[\psi(x+1) - \psi(x-1)]$ and summing upon x we obtain (using the periodicity of ψ, i.e. $\psi(N+1) = \psi(N)$, $\psi(N-1) = \psi(-1)$)

$$\sum_x e^{-q(x)}[\psi^2(x+1) - \psi^2(x-1)] = 0$$

or, equivalently,

$$\sum_x (e^{-q(x-1)} - e^{-q(x+1)})\psi^2(x) = 0.$$

Using (8.37) we reinstate subscripts and dependence on t, and we have

$$\sum_x \left(\exp\{-[q_t(x) + q_t(x-1)]\} - \exp\{-[q_t(x+1) + q_t(x)]\}\right)\phi_j^2(x) = 0 \quad (8.39)$$

for every j.

Combining (8.39) with (8.36), we are led to the following result. If

$$\frac{dq_t(x)}{dt} = \exp\{-[q_t(x) + q_t(x-1)]\} - \exp\{-[q_t(x+1) + q_t(x)]\} \quad (8.40)$$

for $x = 1, 2, \ldots, N$ then the eigenvalues of the matrix $Q(t)$ remain unchanged as t varies.

Set
$$R_t(x) = q_t(x) + q_t(x+1) \tag{8.41}$$

and note that (8.40) implies that
$$\frac{dR_t(x)}{dt} = \exp[-R_t(x-1)] - \exp[-R_t(x+1)]. \tag{8.42}$$

If we introduce quantities $W_t(x)$ by the formula
$$W_t(x) = R_t(x) + R_t(x+1), \tag{8.43}$$

we find easily that
$$\frac{d^2 W_t(x)}{dt^2} = 2\exp[-W_t(x)] - \exp[-W_t(x-2)] - \exp[-W_t(x+2)], \tag{8.44}$$

and setting $W_t(2m) = r_m(t)$ (8.44) becomes
$$\frac{d^2 r_m(t)}{dt^2} = 2\exp[r_m(t)] - \exp[-r_{m-1}(t)] - \exp[-r_{m+1}(t)]. \tag{8.45}$$

These are the equations of motion of the so-called Toda lattice, and the eigenvalues of the matrix $Q(t)$ are related to the constants of motion which have been recently discussed by Flaschka [5] and Hénon [6].

The precise relation between the approach of this section and that of Flaschka will be discussed elsewhere. Here we merely wanted to point out that a discrete analogue of the development of §§ 8.3 and 8.4 is related to a very interesting non-linear Hamiltonian system.

Acknowledgements

This work has been supported in part by AFOSR grant 72-218, and in part by NSF grant 36418X1.

References

1. BUSLAEV, V. S. and FADEEV, L. D. On trace formulas for singular Sturm–Liouville operators. *Dokl. Akad. nauk. SSSR* **132**, 13–16 (in Russian) (1960).
2. GARDNER, C. S., GREENE, J. M., KRUSKAL, M. D., and MINNA, R. M. Method for solving the Konteweg–de Vries equations. *Phys. Rev. Lett.* **19**, 1095–7 (1967).
3. MELNIKOFF, A. The existence of unbounded solutions of the Konteweg–de Vries equation. *Commun. pure appl. Math.* **25**, 407–32 (1972).
4. KAC, M. and VAN MOERBEKE, P. On some isospectral second-order differential operators. *Proc. natn. Acad. Sci. U.S.A.* (To appear.)
5. FLASCHKA, H. The Toda lattice I. Existence of integrals. *Phys. Rev. B* **9**, 1924–5 (1974).
6. HENON, M. Integrals of the Toda lattice. *Phys. Rev. B* **9**, 1921–3 (1974).

9. How to make a heat bath

J. T. Lewis and L. C. Thomas

9.1. Introduction

The probabilistic treatment of Brownian motion began in 1905 with the fundamental work of Einstein ([3], [4]) and Smoluchowski [14]. In their theories the position coordinate $q(t)$ of a Brownian particle of mass m at time t is a random variable and $q(t) - q(0)$ has a Gaussian distribution with mean zero and variance $\sigma^2 |t|$. The effect of the surrounding medium on the motion of the particle is described by two parameters: a temperature T and a frictional constant β. In terms of these parameters the Einstein–Smoluchowski theory gives

$$\sigma^2 = \frac{2kT}{m\beta}.$$

Here k is what is known as Boltzmann's constant. Boltzmann himself always used the gas constant R, and it was Planck who introduced $k = R/N$, where N is Avogadro's number. We mention this because Perrin [15] in a remarkable series of difficult experiments used a determination of σ^2 to estimate N. This brilliant success of the Einstein–Smoluchowski theory tended to distract attention from its highly idealized character and its remoteness from Newtonian mechanics.

Ornstein and Uhlenbeck [20] were the first to provide a dynamical theory of Brownian motion, an idea first sketched by Langevin [9]. They began with Newton's equation of motion

$$p = m \frac{d}{dt} q,$$

$$\frac{d}{dt} p = F. \tag{9.1}$$

They regarded the force F exerted on the particle by the surrounding medium as being made up of two parts: a systematic part $-\beta p$ which acts as a dynamical friction, and a fluctuating random part $mE(t)$ which represents the random impulses which the medium inflicts on the particle. The statistical assumptions which are made about the $E(t)$ are that they have Gaussian distributions with mean zero and that the covariance is given by

$$\langle E(t) E(t+\tau) \rangle = \sigma_0^2 \delta(\tau). \tag{9.2}$$

The resulting equation for the velocity $v(t)$ is the Langevin equation

$$\frac{d}{dt} v(t) + \beta v(t) = E(t). \tag{9.3}$$

On the basis of this equation they found the covariance $\langle u(t)u(t+\tau)\rangle$, where

$$u(t) = v(t) - \langle v(t)\rangle, \tag{9.4}$$

to be given for $\tau > 0$ by

$$\langle u(t)u(t+\tau)\rangle = \frac{\sigma_0^2}{2\beta} e^{-\beta(2t+\tau)}(e^{2\beta t} - 1), \tag{9.5}$$

so that no matter what the initial distribution the limiting distribution has variance $\sigma_0^2/2\beta$. Since the mean satisfies

$$\langle v(t)\rangle = e^{-\beta t}\langle v(0)\rangle, \tag{9.6}$$

the mean value of the kinetic energy satisfies

$$\lim_{t\to\infty} \langle \tfrac{1}{2} mv(t)^2 \rangle = \frac{m\sigma_0^2}{4\beta}. \tag{9.7}$$

But according to the law of the equipartition of energy the mean value of the kinetic energy of a particle in equilibrium with a heat bath at temperature T is $kT/2$. Thus σ_0^2 is put equal to $2kT\beta/m$. On the other hand, we have

$$q(t) - q(0) = \int_0^t v(s)\, ds, \tag{9.8}$$

and it follows that

$$\langle [q(t) - q(0)]^2 \rangle \sim \frac{\sigma_0^2}{\beta^2}|t|, \quad t\to\infty.$$

This agrees with the Einstein–Smoluchowski result when we substitute for σ_0^2.

Mathematicians and physicists both felt uneasy about the Ornstein–Uhlenbeck derivation; mathematicians, because the solution of the 'differential equation' turned out to be non-differentiable; physicists, because, although Newton's equations were used for the particle, the force exerted by the surrounding medium was still treated phenomenologically. The mathematical unease was rapidly relieved by Doob[1] who showed that the re-interpretation of the Langevin equation as an integral equation made possible the use of rigorous methods. In this instance the use of rigour not only removes a source of embarrassment but even simplifies the formal work. We give a brief summary of this in the next section.

The physicists' unease goes deeper. Ideally it should be possible to derive the behaviour of the Brownian particle from the laws of dynamics alone applied to the system consisting of the particles of the medium together with the Brownian particle. That is an ambitious programme. A more modest aim is to provide a mechanical model of the medium and to show that the laws of

How to make a heat bath

dynamics together with the methods of equilibrium statistical mechanics applied to this model yield the required behaviour of the Brownian particle. There is now a vast literature on this subject. The picture which is emerging is that there is a very special model, the Ford–Kac–Mazur model [5], which has some degree of uniqueness and which gives the required behaviour. Many of the other models which have been proposed themselves become the Ford–Kac–Mazur model in the weak-coupling limit (see [7] and [11]).

In the work reported here we show that specifying a Langevin equation involves giving a semigroup of contractions, and that finding the stationary solution of that Langevin equation amounts to finding the unitary dilation of the semigroup of contractions. We show that the Ford–Kac–Mazur model can be regarded as a physical interpretation of the mathematical construction of a unitary dilation. Having done this we are able to show the equivalence of the Ford–Kac–Mazur model with a string model which we have proposed [19].

9.2. Linear stochastic differential equations

It is important for what follows to understand Doob's way out, which we mentioned above. When $E(t)$ is the derivative of a function $w(t)$ the differential equation

$$\frac{dv}{dt} = -\beta v + \frac{dw}{dt} \tag{9.9}$$

is equivalent to the integral equation

$$v(t) - v(t_0) = -\beta \int_{t_0}^{t} v(s)\, ds + w(t) - w(t_0). \tag{9.10}$$

This integral equation continues to make sense even when w is not differentiable. Doob's proposal is to take (9.10) in place of (9.1) as the Langevin equation. In place of condition (9.2) on the $E(t)$ he now requires of the $w(t)$ that they be random variables having mean zero and covariance

$$\langle w(s)w(t) \rangle = \sigma_0^2 (s \wedge t), \tag{9.11}$$

where $s \wedge t$ denotes the minimum of s and t. In other words, $w(t)$ is a Wiener process, just as the Einstein–Smoluchowski $q(t)$ turned out to be. The Langevin equation uses the continuous but non-differentiable $w(t)$ to model the impulses on the Brownian particle from the surrounding medium. If $w(t)$ were differentiable the unique solution of (9.9) satisfying the initial condition $v(0) = v_0$ would be

$$v(t) = e^{-\beta t} v_0 + \int_0^t e^{-\beta(t-s)} w'(s)\, ds.$$

This suggests the solution

$$v(t) = e^{-\beta t}v_0 + \int_0^t e^{-\beta(t-s)}w(ds) \qquad (9.12)$$

for (9.10). There are two obstacles: the integral does not make sense, and there is no way of verifying that (9.12) satisfies (9.10). They are both overcome by the Wiener integral. For each interval $\Delta = (a, b]$ define $w(\Delta)$ by

$$w(\Delta) = w(b) - w(a). \qquad (9.13)$$

Then

$$\langle w(\Delta) \rangle F = 0 \qquad (9.14)$$

and

$$\langle w(\Delta)w(\Delta') \rangle = \sigma_0^2 |\Delta_1 \cap \Delta_2|,$$

where $|\Delta|$ denotes the Lebesgue measure of Δ, as is easily verified using (9.11). Doob [2] made use of property (9.14) to define a random variable

$$w[f] = \int_{-\infty}^{\infty} f(t)w(dt) \qquad (9.15)$$

corresponding to each f in $L^2(\mathbf{R})$. The relationship of $w[f]$ to f has the properties we would expect of an integral; in particular, an 'integration-by-parts' formula holds and we use this to verify that (9.12) satisfies (9.10). Up to this point we have made no mention of the probability space on which the random variables are supposed to live. For example, $w(t)$ is really a function $w(t)(\omega)$ on a space Ω on which a probability measure $\mathbf{P}$ is given so that

$$\langle w(t)w(t+\tau) \rangle = \int_\Omega w(t)(\omega)w(t+\tau)(\omega)\mathbf{P}(d\omega). \qquad (9.16)$$

We are now going to make use of the fact that random variables with zero mean and finite variance can be regarded as elements of a Hilbert space $L^2(\Omega, \mathbf{P})$ with inner product given by the covariance

$$(X, Y) = \langle XY \rangle = \int_\Omega X(\omega)Y(\omega)\mathbf{P}(d\omega). \qquad (9.17)$$

When f is a simple function

$$f(t) = \sum_{i=1}^n c_i \chi_{\Delta_i}(t), \qquad (9.18)$$

$w[f]$ is defined to be

$$w[f] = \sum_{i=1}^n c_i w(\Delta_i), \qquad (9.19)$$

so that

$$(w[f], w[f]) = \langle w[f]^2 \rangle = \sigma_0^2 \int_{-\infty}^{\infty} f(t)^2 \, dt = \|f\|^2. \tag{9.20}$$

Thus $f \mapsto w[f]$ is an isometry from the simple functions in $L^2(\mathbf{R}, \sigma_0^2 \, dt)$ into $L^2(\Omega, \mathbf{P})$. Since the simple functions are dense in $L^2(\mathbf{R}, \sigma_0^2 \, dt)$ it extends by continuity to a unique isometry, also denoted by $f \mapsto w[f]$, from $L^2(\mathbf{R}, \sigma_0^2 \, dt)$ into $L^2(\Omega, \mathbf{P})$. The random variable $w[f]$ so defined is written as an integral

$$w[f] = \int_{-\infty}^{\infty} f(t) w(dt). \tag{9.21}$$

The 'integration-by-parts' formula

$$\int_{-\infty}^{\infty} f(t) w(dt) = -\int_{-\infty}^{\infty} w(t) \, df(t) \tag{9.22}$$

is valid whenever f is a function of bounded variation vanishing outside a finite interval. The details may be found in [14]. The formula

$$\langle w[f] w[g] \rangle = \sigma_0^2 \int_{-\infty}^{\infty} f(t) g(t) \, dt, \tag{9.23}$$

which follows from (9.20) is of great use in the calculation of covariances. For example, using it in conjunction with (9.12) the result (9.5) is easily checked.

9.3. Vector-valued processes

So far we have only considered scalar-valued processes. We need to consider vector-valued (multivariate) processes. For example, Ornstein and Uhlenbeck were able to treat the motion of a harmonically bound Brownian particle. We have in place of (9.3)

$$\frac{dq}{dt} = v,$$

$$\frac{dv}{dt} = -\beta v - \omega_0^2 q + E(t). \tag{9.24}$$

Now we could regard the motion of a point $(q(t), v(t))$ in phase space as a vector-valued stochastic process, but it is more advantageous to adopt the viewpoint of Kolmogorov and define the process X_t as a family of linear mappings from a vector space M into a space $L^2(\Omega, \mathbf{P})$ of random variables. In our example the vector space is $\mathbf{R}^2$; we choose an orthogonal pair of vectors $\{e_1, e_2\}$ in $\mathbf{R}^2$ and define X_t so that

$$X_t e_1 = q(t),$$
$$X_t e_2 = v(t). \tag{9.25}$$

Suppose that the mean of X_t is zero:

$$\langle X_t m \rangle = 0 \quad \text{for all} \quad m \in M.$$

Then the covariance $\langle X_t m X_s m' \rangle$ is a bilinear function on $M \times M$ and we can define a linear operator $R(s, t)$ on M such that

$$\langle X_t m X_s m' \rangle = (m, R(s, t) m')_M. \tag{9.26}$$

We say that X_t is stationary if $R(s, t)$ depends only on the difference $s - t$. To write (9.24) in a way which generalizes (9.10) we need a Wiener process. We define ξ_t to be a linear mapping from a vector space N into $L^2(\Omega, \mathbf{P})$ such that for each n in N the random variables $\xi_t n$ have Gaussian distributions with mean zero and covariance

$$\langle \xi_s n \xi_t n' \rangle = \sigma_0^2 (s \wedge t)(n, n')_N. \tag{9.27}$$

In our example we choose N to be $\mathbf{R}^1$. To complete the description we choose two linear operators $A: M \to N$ and $B: M \to M$; in our example they are given by

$$A = e_1 \otimes \bar{e}_1 \quad \text{and} \quad B = e_2 \otimes \bar{e}_1 - \omega_0^2 e_1 \otimes \bar{e}_2 - \beta e_2 \otimes \bar{e}_2.$$

Then our Langevin equation is taken to be

$$X_t m - X_s m = \int_s^t X_u Bm \, du + \xi_t Am - \xi_s Am, \tag{9.28}$$

which we write formally as

$$dX_t = X_t B \, dt + \xi(dt) A. \tag{9.29}$$

Applying this to our example we obtain

$$dq(t) = v(t) \, dt,$$
$$dv(t) = -\omega_0^2 q(t) \, dt - \beta v(t) \, dt + \xi(dt) e_2. \tag{9.30}$$

To write the solution of (9.28) we need a generalization of the Wiener integral described in § 9.2. We put

$$\xi(\Delta) n = \xi_b n - \xi_a n, \tag{9.31}$$

where $\Delta = (a, b]$, and, using (9.27) we find that

$$\langle \xi(\Delta) n \rangle = 0,$$
$$\langle \xi(\Delta) n \xi(\Delta') n' \rangle = \sigma_0^2 |\Delta \cap \Delta'| (n, n')_N. \tag{9.32}$$

As before, we can define for each f in the space $L^2(\mathbf{R}; N)$ of square-integrable N-valued functions the random variable

$$\xi[f] = \int_{-\infty}^{\infty} \xi(dt) f(t) \tag{9.33}$$

in such a way that

$$\langle \xi[f] \rangle = 0,$$
$$\langle \xi[f]\xi[g] \rangle = \sigma_0^2 \int_{-\infty}^{\infty} (f(t), g(t))_N \, dt. \tag{9.34}$$

An 'integration-by-parts' formula

$$\int_{-\infty}^{\infty} \xi(dt) f(t) = -\int_{-\infty}^{\infty} \xi_t \, df(t) \tag{9.35}$$

holds for suitable f (see [13] for details).

In our example we find that A and B satisfy the following conditions:

(i) for each m in M

$$\lim_{t \to \infty} e^{Bt} m = 0; \tag{9.36}$$

(ii) $$\int_0^{\infty} \|A \, e^{Bt} m\|^2 \, dt < \infty; \tag{9.37}$$

(iii) $$\int_0^{\infty} \|A \, \dot{e}^{Bt} m\|^2 \, dt = 0 \tag{9.38}$$

if and only if $m = 0$;

In general we say that

$$dX_t = X_t B \, dt + \xi(dt) A \tag{9.39}$$

is a Langevin equation if A and B are such that (9.36)–(9.38) hold.

Operating formally with (9.39) we are led to try

$$X_t m = \int_{-\infty}^{t} \xi(ds) A \, e^{(t-s)B} m \tag{9.40}$$

as a solution. Putting

$$\tilde{m}(s) = \begin{cases} A \, e^{-Bs} m, & s \leq 0 \\ 0, & s > 0, \end{cases} \tag{9.41}$$

and

$$(T_t\tilde{m})(s) = \tilde{m}(s-t), \tag{9.42}$$

we can write

$$X_t m = \xi[T_t\tilde{m}]. \tag{9.43}$$

In this form it is easy to check that $X_t m$ has mean zero and finite variance (and hence that it belongs to $L^2(\Omega, \mathbf{P})$) since from (9.34) and (9.37) we have

$$\langle \xi[T_t m]\rangle = 0,$$

$$\langle \xi[T_t m]^2\rangle = \sigma_0^2 \int_0^\infty \|A\,e^{Bs}m\|^2\,ds < \infty. \tag{9.44}$$

Using (9.35) we check that (9.40) satisfies (9.28). It follows from (9.38) that there exists a strictly positive operator R on M such that

$$\langle \xi[T_t m]^2\rangle = (m, Rm) \tag{9.45}$$

and hence that

$$\langle X_t m X_t m'\rangle = (m, Rm'). \tag{9.46}$$

It follows that

$$\langle X_t m X_{t'} m'\rangle = \sigma_0^2 \int_{-\infty}^{t\wedge t'} (A\,e^{B(t-s)}m,\, A\,e^{B(t'-s)}m')\,ds$$

$$= \begin{cases} (m,\, R\,e^{B(t'-t)}m'), & t \le t' \\ (m,\, e^{B^*(t-t')}Rm'), & t' \le t \end{cases} \tag{9.47}$$

In our example we find that, with respect to the basis $\{e_1, e_2\}$, the matrix of e^{Bt} is

$$\begin{bmatrix} e^{-\beta t/2}\left(\cos\omega_1 t + \dfrac{B}{2\omega_1}\sin\omega_1 t\right) & -\dfrac{\omega_0^2}{\omega_1}\sin\omega_1 t\, e^{-\beta t/2} \\ \dfrac{1}{\omega_1}\sin\omega_1 t\, e^{-\beta t/2} & e^{-\beta t/2}\left(\cos\omega_1 t - \dfrac{\beta}{2\omega_1}\sin\omega_1 t\right) \end{bmatrix}$$

and that of R is

$$\begin{bmatrix} \sigma_0^2/2\beta\omega_0^2 & 0 \\ 0 & \sigma_0^2/2\beta \end{bmatrix}.$$

Thus we recover the Ornstein and Uhlenbeck result [21] for the correlation matrix,

$$\begin{bmatrix} \langle q(t)q(t+\tau)\rangle & \langle q(t)v(t+\tau)\rangle \\ \langle v(t)q(t+\tau)\rangle & \langle v(t)v(t+\tau)\rangle \end{bmatrix}$$

$$= \frac{\sigma_0^2}{2\beta} e^{-\beta\tau/2} \begin{bmatrix} \omega_0^{-2}\left(\cos\omega_1\tau + \frac{\beta}{2\omega_1}\sin\omega_1\tau\right) & -\omega_1^{-1}\sin\omega_1\tau \\ \omega_1^{-1}\sin\omega_1\tau & \left(\cos\omega_1\tau - \frac{\beta}{2\omega_1}\sin\omega_1\tau\right) \end{bmatrix}. \quad (9.48)$$

We note that the mean kinetic energy is given by

$$\left\langle \frac{m}{2} v(t)^2 \right\rangle = \frac{m\sigma_0^2}{4\beta}, \quad (9.49)$$

and the mean total energy by

$$\left\langle \frac{m}{2}(v(t)^2 + \omega_0^2 q(t)^2) \right\rangle = \frac{m\sigma_0^2}{2\beta}. \quad (9.50)$$

As before we take

$$\sigma_0^2 = kT\frac{2\beta}{m}.$$

9.4. Dilating semigroups of contractions

In the last section we used the Ornstein–Uhlenbeck example of the Brownian motion of a harmonically bound particle to motivate the definition of a Langevin equation for a vector-valued process X_t; its specification involved a semigroup e^{Bt}, $t \geq 0$, of operators. We found that the stationary solution $X_t \tilde{m}$ of

$$dX_t = X_t B \, dt + \xi(dt) A \quad (9.51)$$

is given by

$$X_t \tilde{m} = \xi[T_t \tilde{m}], \quad (9.52)$$

where $\tilde{m}$ belongs to $L^2(\mathbf{R}; N)$, ξ is a linear mapping from $L^2(\mathbf{R}; N)$ into $L^2(\Omega, \mathbf{P})$ such that

$$\langle \xi[f]^2 \rangle = kT \frac{2\beta}{m} \int_{-\infty}^{\infty} \|f(t)\|_N^2 \, dt \quad (9.53)$$

and T_t is a one-parameter group of unitary operators on $L^2(\mathbf{R}; N)$. We now explore the relationship between $\exp(Bt)$ and T_t.

Suppose we are given a strongly continuous semigroup $\{S_t : t \geq 0, S_0 = 1\}$ of contraction operators ($\|S_t\| \leq 1$) on a Hilbert space M; can we find an isometry J of M into a Hilbert space $\hbar$ and a group T_t of unitary operators on $\hbar$ such that for all $m \in M$ and all $t \geq 0$

$$JS_t m = PT_t Jm, \tag{9.54}$$

where P is the orthogonal projection of $\hbar$ onto the image $\tilde{M}$ of M under J? The question was answered in the affirmative by Sz-Nagy [17]: the group T_t is called a unitary dilation of the semigroup S_t. In the case when S_t tends strongly to zero, i.e. when, for all $m \in M$,

$$\lim_{t \to \infty} S_t m = 0, \tag{9.55}$$

an explicit construction has been given independently by Lax and Phillips [10] and Sz-Nagy and Foias [18]. Let B be the generator of S_t, so that $S_t = \exp Bt$. Then there exists a linear mapping A from the domain of B into a Hilbert space N and an isometry J of M into $L^2(\mathbf{R}; N)$ given on the domain of B by

$$\tilde{m}(s) = \begin{cases} A e^{-Bs} m, & s \leq 0 \\ 0, & s > 0. \end{cases} \tag{9.56}$$

Let P be the projection defined by

$$(Pf)(s) = \chi_{(-\infty, 0]}(s) f(s) \tag{9.57}$$

and let T_t be the group of unitary operators on $L^2(\mathbf{R}; N)$ given by

$$(T_t f)(s) = f(s - t). \tag{9.58}$$

Then

$$(JS_t m)(s) = PT_t \tilde{m}(s). \tag{9.59}$$

We sketch their construction.

The domain $\mathcal{D}(B)$ of B is dense in M and since S_t is a contraction, the form $[.,.]$ defined on $\mathcal{D}(B)$ by

$$[m, m] = -(k, Bk)_M - (Bk, k)_M \tag{9.60}$$

is non-negative. Let $\mathcal{D}(B)_0$ be the set of vectors in $\mathcal{D}(B)$ for which $[m, m]$ is zero. The form $[.,.]$ induces an inner product $(.,.)_N$ on $\mathcal{D}(B)/\mathcal{D}(B)_0$, and we define N to be the completion of $\mathcal{D}(B)/\mathcal{D}(B)_0$ in the associated norm.

Let A be the quotient map of $\mathcal{D}(B)$ into N. For m in $\mathcal{D}(B)$ define $\tilde{m}$ by (9.56). Then

$$\int_{-\infty}^{\infty} \|\tilde{m}(s)\|_N^2 \, ds = -\int_{-\infty}^{0} [(B\,e^{-Bs}m, e^{-Bs})_M + (e^{-Bs}m, B\,e^{-Bs}m)_M]\, ds$$

$$= \int_{-\infty}^{0} \frac{d}{ds}\|e^{-Bs}m\|_M^2 \, ds$$

$$= \|m\|_M^2, \tag{9.61}$$

since

$$\lim_{s\to\infty} \|e^{Bs}m\| = 0.$$

Hence $m \mapsto \tilde{m}$ is an isometry. We see that (9.59) holds since

$$J\,e^{Bt}m = \begin{cases} A\,e^{-Bs}\,e^{Bt}m, & s \leq 0 \\ 0, & s > 0 \end{cases} \tag{9.62}$$

$$= \chi_{(-\infty,0]}(s)\,(T_t\tilde{m})(s).$$

This looks like the construction used in the solution (9.52) of the Langevin equation. To make the correspondence precise we must pay more attention to norms than we did in § 9.3. In the Ornstein–Uhlenbeck example we used the usual norm in $\mathbf{R}^2$; with respect to this $\exp(Bt)$ may fail to be a contraction. We can put this right by making use of the energy norm, as we shall show when we consider the Ford–Kac–Mazur model in the next section. Then the correspondence is exact.

9.5. The Ford–Kac–Mazur model

We are going to analyse the model of a heat bath given by Ford, Kac, and Mazur [5]. In summarizing their work we cannot do better than follow closely the clear expository account given by Kac [6].

Consider a dynamical system consisting of the Brownian particle, whose motion is governed by the Hamiltonian

$$H_0 = \frac{1}{2m}p_0^2 + V(q_0), \tag{9.63}$$

coupled to a dynamical system of a large number $2N$ of degrees of freedom (the heat bath), whose motion is governed by the Hamiltonian

$$H_N = \sum_{\{j:\,0<|j|\leq N\}} \frac{1}{2m_j}p_j^2 + V(q_{-N},\ldots,q_{-1},q_1,\ldots,q_N), \tag{9.64}$$

by a coupling
$$H_c = W(q_{-N}, \ldots, q_{-1}, q_0, q_1, \ldots, q_N). \tag{9.65}$$

The total Hamiltonian
$$H = H_0 + H_c + H_N \tag{9.66}$$

gives rise to equations of motion which when solved will yield for the motion of the Brownian particle

$$\begin{aligned} p_0(t) &= F_t(p_{-N}(0), \ldots, p_N(0); q_{-N}(0), \ldots, q_N(0)), \\ q_0(t) &= G_t(p_{-N}(0), \ldots, p_N(0); q_{-N}(0), \ldots, q_N(0)). \end{aligned} \tag{9.67}$$

At $t=0$ the heat bath is assumed to be in equilibrium at temperature T, i.e., the initial values of the momenta and coordinates $(p_{-N}(0), \ldots, p_N(0); q_{-N}(0), \ldots, q_N(0))$ are assumed to be distributed according to the canonical distribution at temperature T: their joint probability distribution is proportional to

$$\exp\left[-\frac{1}{kT} H_N(p_{-N}(0), \ldots, p_N(0); q_{-N}(0), \ldots, q_N(0))\right]. \tag{9.68}$$

In this way $(p_0(t), q_0(t))$ becomes a stochastic process in the (p_0, q_0) phase plane. We are faced with the following.

Problem 1. Can H_N and H_c be chosen so that the statistical description of the process $(p_0(t), q_0(t))$ agrees, to a good approximation, with that derived using Langevin's equation?

If we can solve this we are then faced with the following.

Problem 2. Can H_N and H_c be chosen in a realistic fashion?

Ford, Kac, and Mazur begin by solving a modified version of Problem 1. They put $V(q_0) \equiv 0$ in (9.63) and assume that at $t=0$ the whole system is in equilibrium at temperature T. That is, the joint probability distribution of $(p_{-N}(0), \ldots, p_0(0), \ldots, p_N(0); q_{-N}(0), \ldots, q_0(0), \ldots, q_N(0))$ is proportional to

$$\exp\left[-\frac{1}{kT} H(p_{-N}(0), \ldots, p_0(0), \ldots, p_N(0); q_{-N}(0), \ldots, q_0(0), \ldots, q_N(0))\right]. \tag{9.69}$$

They choose H to be a quadratic

$$H = \frac{1}{2} \sum_{j=-N}^{j=N} p_j^2 + \frac{1}{2} \sum_{j,k} (\Omega^2)_{j,k} q_j q_k, \tag{9.70}$$

where Ω^2 is positive-definite, so that the distribution (9.69) is Gaussian. In fact they choose Ω^2 to be given by

$$(\Omega^2)_{j,k} = \frac{1}{2\pi}\int_{-\pi}^{\pi} f^2(\theta)\, e^{i(j-k)\theta}\, d\theta. \quad (9.71)$$

Because the Hamiltonian is quadratic the equations of motion can be solved explicitly and the covariances calculated. For example, they find that

$$\langle p_0(t)p_0(t+\tau)\rangle = kT(\cos\tau\Omega)_{00}. \quad (9.72)$$

In § 9.1 we saw that for the Brownian motion of a free particle

$$\langle p_0(t)p_0(t+\tau)\rangle = kT\, e^{-\beta|\tau|}, \quad (9.73)$$

so we would like to choose Ω so that

$$(\cos\Omega\tau)_{00} = e^{-\beta|\tau|}. \quad (9.74)$$

This cannot be done with N finite but it can be done for an infinite heat bath if we choose $f(\theta)$ to be given by

$$f(\theta) = \beta\tan\frac{\theta}{2}, \quad (9.75)$$

so that

$$(\cos\Omega\tau)_{j,k} = \frac{1}{2\pi}\int_{-\pi}^{\pi}\cos\left(\beta\tau\tan\frac{\theta}{2}\right) e^{i(j-k)\theta}\, d\theta. \quad (9.76)$$

The fact that $f(\theta)$ is not integrable on $(-\pi, \pi)$ causes difficulties if we insist, as Ford, Kac, and Mazur do, on the coupled oscillator interpretation of the equations of motion. For this reason they are forced to employ a cut-off

$$f(\theta) = \begin{cases} \beta\tan\dfrac{\theta}{2}, & |\theta|<\theta_0 \\ 0, & \theta_0<|\theta|<\pi. \end{cases} \quad (9.77)$$

Then in the limit $\theta_0 \to \pi$, following the limit $n \to \infty$, they obtain (9.73).

If we do not insist on the coupled oscillator interpretation we can still retain Hamiltonian equations of motion for an infinite system, but we have to be a little more sophisticated in describing the stochastic process. There are two independent features of the Ford–Kac–Mazur model:
  (i) the introduction of the dissipative mechanism and
  (ii) the specification of the stochastic process.
In our analysis we separate them.

With the simplified Hamiltonian (9.70) the equations of motion become

$$\frac{dq}{dt} = p, \qquad \frac{dp}{dt} = -\Omega^2 q, \tag{9.78}$$

where p and q are vectors in $\mathbf{R}^{2N+1}$. It is convenient to introduce the vector

$$z = (p + i\Omega q) \tag{9.79}$$

in $\mathbf{C}^{2N+1}$ and then the equation of motion becomes

$$\frac{dz}{dt} = i\Omega z. \tag{9.80}$$

With respect to the standard inner product

$$(z, z') = \sum_{j=-N}^{j=N} \bar{z}_j z'_j \tag{9.81}$$

in $\mathbf{C}^{2N+1}$ the operator Ω is self-adjoint and

$$U_t = e^{it\Omega} \tag{9.82}$$

is a one-parameter group of unitary operators.
Notice that

$$\tfrac{1}{2}\|z\|^2 = \tfrac{1}{2}\bigg(\sum_{j=-N}^{j=N} p_j^2 + \sum_{j,k} (\Omega^2)_{j,k} q_j q_k\bigg) \tag{9.83}$$

is the total energy.

In this form the generalization to a Hamiltonian system with an infinite number of degrees of freedom is straightforward. We take the complexified phase space to be the sequence space l^2 with the standard inner product

$$(z, z') = \sum_{-\infty}^{\infty} \bar{z}_j z'_j \tag{9.84}$$

and standard basis $\{\ldots, e_{-1}, e_0, e_1, \ldots\}$, where

$$(e_k)_j = \delta_{jk}. \tag{9.85}$$

Then we interpret $\tfrac{1}{2}\|z\|^2$ as the total energy associated with the vector z and specify the dynamics by giving a one-parameter unitary group U_t. The generator Ω of U_t may be an unbounded operator but that is unimportant. The (p_0, q_0) phase plane now corresponds to the one-dimensional subspace spanned by e_0, and we let P_0 denote the projection on it, given by

$$P_0 z = (e_0, z) e_0. \tag{9.86}$$

What we want is a dissipative mechanism by which the one-dimensional system is embedded in an infinite conservative system in such a way that the restriction to the one-dimensional system produces frictional damping. The problem of finding such a system amounts to the following.

Find a one-parameter unitary group U_t on l^2 such that for $t \geq 0$

$$P_0 U_t P_0 = e^{-\beta t} P_0.$$

The solution to this is contained in the Ford–Kac–Mazur model. We make use of the isomorphism between $L^2(-\pi, \pi)$ and l^2 given by the Fourier transform F,

$$(Fg)_n = \int_{-\pi}^{\pi} \frac{e^{-in\theta}}{(2\pi)^{\frac{1}{2}}} g(\theta) \, d\theta. \tag{9.87}$$

Under this isomorphism U_t corresponds to the unitary group $\check{U}_t$ on $L^2(-\pi, \pi)$ given by

$$(\check{U}_t g)(\theta) = \exp\left(-it\beta \tan \frac{\theta}{2}\right) g(\theta), \tag{9.88}$$

with generator $\check{\Omega}$ given by

$$(\check{\Omega} g)(\theta) = -\beta \tan \frac{\theta}{2} g(\theta). \tag{9.89}$$

The projection P_0 corresponds to $\check{P}_0$ given by

$$(\check{P}_0 g)(\theta) = \int_{-\pi}^{\pi} g(\theta) \frac{d\theta}{2\pi}, \tag{9.90}$$

so that

$$\check{P}_0 \check{U}_t \check{P}_0 = \int_{-\pi}^{\pi} \exp\left(it\beta \tan \frac{\theta}{2}\right) \frac{d\theta}{2\pi} \check{P}_0$$

$$= e^{-\beta |t|} \check{P}_0, \tag{9.91}$$

and hence

$$P_0 U_t P_0 = e^{-\beta |t|} P_0, \tag{9.92}$$

as required.

To see how to specify the stochastic process in the system with an infinite number of degrees of freedom we return to the finite system. Following Ford, Kac, and Mazur we could require that at $t = 0$ the momenta and coordinates be given a joint probability distribution proportional to

$$\exp\left(-\frac{1}{2kT}\|z\|^2\right), \tag{9.93}$$

since $\frac{1}{2}\|z\|^2$ is the total energy.

Equivalently, we could specify the stochastic process by giving a linear mapping ζ from $\mathbf{C}^{2N+1}$ into a space $L^2(\Omega, \mathbf{P})$ of complex-valued random variables, so that for each z in $\mathbf{C}^{2N+1}$ the random variable $\zeta[z]$ has a Gaussian distribution with mean zero and variance

$$\langle |\zeta[z]|^2 \rangle = kT \|z\|^2. \tag{9.94}$$

This second formulation continues to make sense in the case of an infinite number of degrees of freedom. Applying this to the dissipative mechanism already constructed we get the process corresponding to $p_0(t)$ to be

$$Y_t = \zeta[U_t e_0], \tag{9.95}$$

where

$$(e_0)_n = \delta_{0n}. \tag{9.96}$$

Then

$$\langle \bar{Y}_t Y_{t+\tau} \rangle = \langle \overline{\zeta[U_t e_0]} \zeta[U_{t+\tau} e_0] \rangle$$
$$= kT(e_0, U_\tau e_0)$$
$$= kT e^{-\beta|\tau|}, \tag{9.97}$$

in agreement with (9.73).

The problem of constructing a dissipative mechanism, which we have seen is solved by the Ford–Kac–Mazur model, is just a special case of the problem which we discussed in § 9.4. Let us see what the construction of § 9.4 leads to in this case. We have an isometry from the one-dimensional space $P_0 l^2$ into $L^2(\mathbf{R}; \mathbf{C})$ given by

$$e_0 \mapsto \tilde{e}_0(s) = \begin{cases} (2\beta)^{\frac{1}{2}} e^{\beta s}, & s \leq 0 \\ 0, & s > 0. \end{cases} \tag{9.98}$$

The unitary dilation of the semigroup $e^{-\beta t}$, $t \geq 0$, is then given by T_t, where

$$(T_t g)(s) = g(s - t). \tag{9.99}$$

As we might expect, T_t and U_t are unitarily equivalent. Let $\mathscr{F}: L^2(\mathbf{R}) \to L^2(\mathbf{R})$ be the Fourier transform, under which T_t goes into multiplication by $\exp(-ikt)$, so that

$$(\mathscr{F}\tilde{e}_0)(k) = \left(\frac{\beta}{\pi}\right)^{\frac{1}{2}} \frac{1}{\beta - ik}. \tag{9.100}$$

Let $V: L^2(\mathbf{R}) \to L^2(-\pi, \pi)$ be the unitary mapping given by

$$(Vf)(\theta) = \left(\frac{\beta}{2}\right)^{\frac{1}{2}} \left(1 - i \tan \frac{\theta}{2}\right) f\left(\beta \tan \frac{\theta}{2}\right), \tag{9.101}$$

under which multiplication by exp(i*k*t) goes into multiplication by

$$\exp[-it\beta \tan\theta/2]$$

and

$$(V\mathscr{F}\tilde{e}_0)(\theta) = \frac{1}{\sqrt{(2\pi)}} = \check{e}_0(\theta). \tag{9.102}$$

Let ξ be the Wiener mapping considered in § 9.3, now acting on $L^2(\mathbf{R}; \mathbf{C})$ with $\sigma_0^2 = kT$. Then the process

$$X_t = \xi[T_t\tilde{e}_0] \tag{9.103}$$

is a Gaussian process with mean zero and covariance

$$\overline{\langle \xi[T_t\tilde{e}_0]\xi[T_{t+\tau}\tilde{e}_0]\rangle} = kT\,e^{-\beta|\tau|}. \tag{9.104}$$

so that X_t and the Ford–Kac–Mazur process Y_t are equivalent.

9.6. A string model

There is a model of a dissipative system which uses a stretched string as the system with an infinite number of degrees of freedom. This was proposed by Lamb [8] in 1900 as a model of radiation damping, and used by us as an alternative to the Ford–Kac–Mazur model (see [12]).

A particle of mass m is constrained to move along the y-axis, and it is attached to a string of density ρ stretched along the positive x-axis at tension F.

The wave velocity c in the string is given by

$$c^2 = F/\rho.$$

Introducing

$$\lambda = m/\rho$$

as a characteristic length and

$$\tau_1 = \lambda/c$$

as a characteristic time we find that the equations of motion of particle and string can be written in dimensionless form

$$u_{tt}(0, t) = u_x(0, t), \tag{9.105}$$

$$u_{tt}(x, t) = u_{xx}(x, t), \quad x > 0. \tag{9.106}$$

We shall make use of the norm $\|u\|$ given by

$$\|u\|^2 = \frac{1}{2}\left\{u_t(0, t)^2 + \int_0^\infty [u_x(s, t)^2 + u_t(s, t)^2]\,ds\right\}, \tag{9.107}$$

which is chosen so that the total energy $E(t)$ is given by
$$E(t) = mc^2 \|u\|^2. \tag{9.108}$$
Using the equation of motion (9.106) we find that
$$\frac{dE}{dt}(t) = mc^2 \left\{ u_t(0, t)u_{tt}(0, t) + \int_0^\infty [u_x(s, t)u_{xt}(s, t) + u_t(s, t)u_{tt}(s, t)] \, ds \right\}$$
$$= mc^2 \left\{ u_t(0, t)u_{tt}(0,t) + \int_0^\infty \frac{\partial}{\partial x}[u_x(s, t)u_t(s, t)] \, ds \right\}$$
$$= mc^2 \{u_t(0, t)u_{tt}(0, t) - u_x(0, t)u_t(0, t)\}.$$
Invoking the boundary condition (9.105) we obtain
$$\frac{dE}{dt}(t) = 0. \tag{9.109}$$
Thus if the initial conditions are given by the vector-valued function $f(x)$ with components
$$f_1(x) = u(x, 0),$$
$$f_2(x) = u_t(x, 0), \tag{9.110}$$
we have
$$E(t) = E(0) = mc^2\|f\|_E^2,$$
where
$$\|f\|_E^2 = \frac{1}{2} \left\{ f_2(0)^2 + \int_0^\infty [f_1'(s)^2 + f_2(s)^2] \, ds \right\}. \tag{9.111}$$
Let h_E be the Hilbert space formed by completing the linear space consisting of those initial data functions f which are infinitely differentiable and vanish outside some bounded interval, in the energy norm (9.111). A solution of (9.106) subject to (9.105) can be regarded as a flow on h_E. If $f(x)$ corresponds to the solution $x, t \mapsto u(x, t)$, then let $f_T(x)$ be the initial data corresponding to the solution $x, t \mapsto u(x, t+T)$. Because of (9.109) the flow $f \mapsto f_T$ is a unitary mapping which we denote by U_T.

The general solution of (9.106) is given by a pair a, b of functions on $\mathbf{R}$,
$$u(x, t) = a(t+x) + b(t-x). \tag{9.112}$$
Because of the boundary condition (9.105) the functions a and b are not independent. From the initial data $a'(s)$ is determined directly for $s > 0$ and $b'(s)$ for $s < 0$,
$$a'(s) = \tfrac{1}{2}(f_1'(s) + f_2(s)), \quad s > 0, \tag{9.113}$$
$$b'(s) = \tfrac{1}{2}(-f_1'(-s) + f_2(-s)), \quad s < 0. \tag{9.114}$$

Putting
$$v(t) = u_t(0, t) \tag{9.115}$$
in the boundary condition (9.105) and using (9.112) we obtain
$$\frac{d}{dt}v(t) + v(t) = 2a'(t), \tag{9.116}$$
or
$$\frac{d}{dt}v(t) - v(t) = -2b'(t). \tag{9.117}$$
Solving (9.116) determines $b'(s)$ for $s \geq 0$,
$$b'(s) = -a'(s) + e^{-s}f_2(0) + 2\int_0^s e^{-(s-u)}a'(u)\,du, \tag{9.118}$$
and solving (9.117) determines $a'(s)$ for $s \leq 0$,
$$a'(s) = -b'(s) + e^s f_2(0) - 2\int_0^s e^{(s-u)}b'(u)\,du. \tag{9.119}$$

Reverting, for the moment, to the non-dimensionless $t' = t\tau_1$ and putting $\beta = 1/\tau_1$, we see that (9.116) becomes
$$\frac{dv}{dt'}(\beta t') + \beta v(\beta t') = 2\frac{d}{dt'}a(\beta t'). \tag{9.120}$$

This shows how the coupling of the particle to the semi-infinite string produces a damping term $-\beta v(\beta t')$ and a driving term $2(d/dt')a(\beta t')$.

We can regard multiplication by $\exp(-t)$ as a semigroup of contraction operators on the two-dimensional real vector space M consisting of the initial data $f(0)$ of the particle at $x = 0$. Solving the wave equation amounts to dilating this semigroup. This is most conveniently seen by passing to the translation representation [10]). We construct a unitary mapping V of h_E onto $L^2(\mathbf{R})$ so that U_t goes into translation to the right by t. Put
$$h(s) = (Vf)(s) = b'(-s), \tag{9.121}$$
where b' is determined from f by (9.114) and (9.118).
Explicitly
$$b'(-s) = \begin{cases} \frac{1}{2}[-f_1'(s) + f_2(s)], & s > 0 \\ -\frac{1}{2}[f_1'(-s) + f_2(-s)] + e^s f_2(0) + \\ \qquad + \int_0^{-s} e^{(s+u)}[f_1'(u) + f_2(u)]\,du, & s \leq 0. \end{cases} \tag{9.122}$$

It follows from (9.112) that
$$(VU_t f)(s) = b'(t-s) = h(s-t). \tag{9.123}$$

It remains to show that
$$\|f\|_E^2 = \int_{-\infty}^{\infty} h(s)^2 \, ds. \tag{9.124}$$

Now
$$\|f\|_E^2 = \frac{1}{2}\left([a'(0)+b'(0)]^2 + \int_0^{\infty} \{[a'(s)-b'((-s))]^2 + [a'(s)+b'(-s)]^2\} \, ds\right)$$
$$= \tfrac{1}{2}[a'(0)+b'(0)]^2 + \int_0^{\infty} a'(s)^2 \, ds + \int_{-\infty}^{0} b'(s)^2 \, ds. \tag{9.125}$$

But the boundary condition (9.105) yields
$$a''(s) + b''(s) = a'(s) - b'(s), \tag{9.126}$$

so that
$$\frac{d}{ds}\{\tfrac{1}{2}[a'(s)+b'(s)]^2\} = a'(s)^2 - b'(s)^2,$$

and so
$$\tfrac{1}{2}[a'(0)+b'(0)]^2 = -\int_0^{\infty} a'(s)^2 \, ds + \int_0^{\infty} b'(s)^2 \, ds. \tag{9.127}$$

Substituting in (9.124) gives
$$\|f\|_E^2 = \int_{-\infty}^{\infty} b'(-s)^2 \, ds. \tag{9.128}$$

Notice that if f is in M,
$$f_1'(s) = 0, \, f_2(s) = 0, \quad s > 0. \tag{9.129}$$

So from (9.122)
$$h(s) = (Vf)(s) = \begin{cases} 0, & s > 0 \\ e^s f_2(0), & s \leq 0. \end{cases} \tag{9.130}$$

We have produced the same dilation of the semigroup as we had in § 9.5. As before, having produced the dissipative structure we specify the stochastic process by Wienerizing,
$$X_t = \xi[VU_t f]. \tag{9.131}$$

It is easy to generalize this model to produce a model of the Brownian motion of a system of coupled oscillators. We content ourselves with the Ornstein and Uhlenbeck example which we considered in § 9.3. We want to model the system (9.24)

$$\frac{dq}{dt} = v,$$

$$\frac{dv}{dt} = -\beta v - \omega_0^2 q + E(t).$$

We take a particle of mass m constrained to move along the y-axis subject to a restoring force $-m\omega_0^2 y$, and attach it to a string of density ρ stretched along the x-axis at tension F. In addition to the characteristic length $\lambda = m/\rho$ and the characteristic time $\tau_1 = \lambda/c$ we have a second characteristic time $\tau_2 = 1/\omega_0$. Hence we have a dimensionless constant

$$\gamma = \tau_2/\tau_1.$$

We consider the case $\gamma < 1$ (the other cases are treated similarly). The equations of motion of particle and string can be put in the form

$$u_{tt}(0, t) + \gamma^2 u(0, t) = u_x(0, t), \tag{9.132}$$

$$u_{tt}(x, t) = u_{xx}(x, t), \quad x > 0, \tag{9.133}$$

where λ and τ_1 have been used to make the equations dimensionless. Put

$$y_1(t) = \gamma u(0, t), \quad y_2(t) = u_t(0, t) \tag{9.134}$$

and

$$\|y\|^2 = y_1(t)^2 + y_2(t)^2, \tag{9.135}$$

so that

$$\frac{mc^2}{2}\|y\|^2 = \frac{mc^2}{2}[u_t(0, t)^2 + \gamma^2 u(0, t)^2] \tag{9.136}$$

is the total energy of the particle in isolation. The equation of motion of the particle becomes

$$\frac{d}{dt}y_1(t) = \gamma y_2(t)$$

$$\frac{d}{dt}y_2(t) = -\gamma y_1(t) - y_2(t) + 2a'(t), \tag{9.137}$$

which can be written

$$\frac{d}{dt}y = Sy - Py + 2a'(t)e_2,$$

where
$$y(t) = y_1(t)e_1 + y_2(t)e_2,$$
$$Py(t) = y_2(t)e_2,$$

and S has matrix

$$\begin{bmatrix} 0 & \gamma \\ -\gamma & 0 \end{bmatrix} \tag{9.138}$$

so that $\exp St$ preserves the norm of y. It follows that $\exp Bt$ is a one-parameter semigroup of contractions, where

$$B = S - P. \tag{9.139}$$

We have to check that $\|e^{Bt}\| \leq 1$ for $t \geq 0$. Putting $x(t) = e^{Bt}x$ we have

$$\frac{d}{dt}\|x(t)\|^2 = ((S-P)x(t), x(t)) + (x(t), (S-P)x(t))$$
$$= -2(x(t), Px(t)) \leq 0,$$

since $(X(t), Sx(t)) = 0$ and P is a positive symmetric operator. Hence

$$\|e^{Bt}x\|^2 - \|x\|^2 = \int_0^t \frac{d}{dt}\|x(t)\|^2 \, dt \leq 0,$$

so that
$$\|e^{Bt}\| \leq 1.$$

The total energy $E(t)$ of particle and string is given by

$$E(t) = \frac{mc^2}{2}\left\{ u_t(0, t)^2 + \gamma^2 u(0, t)^2 + \int_0^\infty [u_x(s, t)^2 + u_t(s, t)^2] \, ds \right\}. \tag{9.140}$$

Introducing

$$\|f\|_E^2 = \frac{1}{2}\left\{ \|y(0)\|^2 + \int_0^\infty [f_1'(s)^2 + f_2(s)^2] \, ds \right\} \tag{9.141}$$

we can again regard the solution of the equations of motion as a flow on the initial data space $\hbar_E$ given by a unitary operator U_t. It turns out that this is a dilation of the contraction semigroup $\exp Bt$ acting on $M = \mathbf{R}^2$.

Eqn (9.137) for the oscillator can be written as

$$\frac{dy}{dt} = By + 2a'(t)e_2. \tag{9.142}$$

As before, we have a' given for $s>0$ by (9.113) and b' given for $s<0$ by (9.124). To get b' for $s>0$ we have to solve (9.143). We have

$$b'(-s) = \begin{cases} \frac{1}{2}[-f_1'(s)+f_2(s)], & s>0 \\ -\frac{1}{2}[f_1'(-s)+f_2(-s)]+[e_2, e^{-Bs}y(0)] + \\ \quad + \int_0^{-s} (e_2, e^{-B(s+u)}e_2)[f_1'(u)+f_2(u)]\,du, & s\leq 0, \end{cases} \quad (9.143)$$

where

$$y(0) = \gamma f_1(0)e_1 + f_2(0)e_2.$$

Again the mapping V of $f(s)$ to $b'(-s)$ given by (9.143) is a unitary mapping of $\hbar_E$ onto $L^2(\mathbf{R})$ under which U_t becomes right translation by t. To check that it preserves norm we have to use (9.142):

$$\tfrac{1}{2}\|y(0)\|^2 = -\frac{1}{2}\int_0^\infty \frac{d}{dt}\|y(t)\|^2\,dt$$

$$= \int_0^\infty y_2(t)[y_2(t)-2a'(t)]\,dt$$

$$= \int_0^\infty b'(t)^2\,dt - \int_0^\infty a'(t)^2\,dt. \quad (9.144)$$

But

$$\frac{1}{2}\int_0^\infty [f_1'(t)^2+f_2(t)^2]\,dt = \int_0^\infty a'(t)^2\,dt + \int_{-\infty}^0 b'(t)^2\,dt \quad (9.145)$$

so that

$$\|f\|_E^2 = \int_{-\infty}^\infty b'(-s)^2\,ds. \quad (9.146)$$

For f in the subspace M corresponding to the initial data of the oscillator we have

$$h(s) = (Vf)(s) = \begin{cases} 0, & s>0 \\ (e_2, e^{-Bs}y(0)), & s\leq 0. \end{cases} \quad (9.147)$$

Using the Wiener mapping introduced in § 9.3 we get the process

$$X_t y(0) = \int_{-\infty}^t \xi(ds)A\,e^{B(t-s)}y(0) \quad (9.148)$$

obtained there, where A is the projection on e_2.

9.7. Lax-Phillips structures

The string models considered in § 9.6 are examples of classical scattering systems whose Hilbert-space structure has been investigated by Lax and Phillips [10]. In [13] we pointed out that stationary solutions of Langevin equations have the same Hilbert-space structure. It is this result which lies behind the analysis presented here. We used in § 9.2 the fact that a space of random variables having zero mean and finite variances can be regarded as a Hilbert space $L^2(\Omega, \mathbf{P})$ with the inner product given by the covariance. Cramer and Kolmogorov showed that it can be profitable to regard a stochastic process as a curve in this Hilbert space. Using the notation of § 9.3, we consider a vector-valued process X_t. For each vector in the vector space M the random variable $X_t m$ traces out a curve as t runs from $-\infty$ to ∞. As m runs through M we get a bunch of curves

$$\{X_t m: -\infty < t < \infty, m \in M\}.$$

The closed subspace $\hbar$ of $L^2(\Omega, \mathbf{P})$ which is spanned by the points on this bunch of curves is called the history of the process and we write it as

$$\hbar = V\{X_t m: -\infty < t < \infty, m \in M\}.$$

We need to consider the closed subspace $\hbar_t$ of $\hbar$ called the history of the process up to time t defined by

$$\hbar_t = V\{X_s m: -\infty < s \leq t, m \in M\}.$$

If the process is stationary the operator U_t defined on $\hbar$ by

$$U_t X_0 m = X_t m \tag{9.149}$$

is unitary. We shall assume that X_t has sufficient continuity properties to make U_t a strongly-continuous one-parameter unitary group. We have

$$U_t \hbar_s = \hbar_{s+t} \tag{9.150}$$

and

$$\bigvee_{-\infty < t < \infty} \hbar_t = \hbar. \tag{9.151}$$

In prediction theory an important class of processes consists of those whose remote past, the intersection of all the $\hbar_t$, is the zero vector alone,

$$\bigwedge_{-\infty < t < \infty} \hbar_t = \{0\}. \tag{9.152}$$

They are called regular (or purely non-deterministic). A regular stationary process has a representation as a moving average (see [13])

$$X_t m = \int_{-\infty}^{t} \xi(\mathrm{d}s) \tilde{m}(s-t). \tag{9.153}$$

We say that $(\hbar, T_t, \mathcal{D})$ is a K-structure (K for Kolmogorov) if T_t is a strongly continuous one-parameter unitary group of operators on $\hbar$ and $\mathcal{D}$ is a closed subspace of $\hbar$ whose image under T_t is denoted by $\mathcal{D}_t$, and
  (i) $\mathcal{D} \subseteq \mathcal{D}_t$ for all $t \geq 0$,
 (ii) $V \mathcal{D}_t = \hbar$,
(iii) $\bigwedge \mathcal{D}_t = \{0\}$.

Evidently X_t is a regular stationary process if and only if $(\hbar, U_t, \hbar_0)$ is a K-structure. We say that $(\hbar, T_t, \mathcal{D}_-, \mathcal{D}_+)$ is an LP-structure if $(\mathcal{D}_-, \mathcal{D}_+)$ is a pair of orthogonal subspaces such that $(\hbar, T_t, \mathcal{D}_-)$ and $(\hbar, T_t^*, \mathcal{D}_+)$ are K-structures. We say that an LP-structure is cyclic if

$$VT_t \mathcal{K} = \hbar, \tag{9.154}$$

where

$$\mathcal{K} = (\mathcal{D}_+ \oplus \mathcal{D}_-)^\perp.$$

Returning to the Langevin equation

$$dX_t = X_t B \, dt + \xi(dt) A, \tag{9.155}$$

where $B: M \to M$ is the generator of a strongly continuous semigroup of contractions tending strongly to zero as $t \to \infty$, and $A: M \to N$ is such that

$$\int_0^\infty \|A \, e^{Bt} m\|_N^2 \, dt = \|m\|_M^2, \tag{9.156}$$

we see that the stationary solution

$$X_t m = \int_{-\infty}^t \xi(ds) A \, e^{B(t-s)} m \tag{9.157}$$

is regular with K-structure $(\hbar, U_t, \hbar_0)$ so $(\hbar, U_t^*, \hbar_0^\perp)$ is also a K-structure. Put $\mathcal{D}_+ = \hbar_0^\perp$, $\mathcal{K} = \{X_0 m : m \in M\}$ and $\mathcal{D}_- = \{\mathcal{D}_+ \oplus \mathcal{K}\}^\perp$; then $(\hbar, U_t, \mathcal{D}_-, \mathcal{D}_+)$ is a cyclic LP-structure. This is proved in [13] together with the following converse.

In a cyclic LP-structure the restriction S_t of T_t to $\mathcal{K}$ is, for $t \geq 0$, a strongly continuous semigroup of contractions tending strongly to zero as $t \to \infty$ (see [10]). Let j be the injection of $\mathcal{K}$ in $\hbar$ and let B be the generator of S_t. Then there exists a Hilbert space N, an operator-valued Wiener process $\xi_t: N \to \hbar$ and a linear mapping $A: \mathcal{D}(B) \to N$ such that the process X_t given by $T_t \circ j$ is the unique regular solution of the Langevin equation

$$dX_t = X_t B \, dt + \xi(dt) A.$$

LP-structures were first observed in classical scattering theory [10].

Consider the example of the wave equation (9.133) subject to the boundary condition (9.132). Then $(\hbar_E, U_t, \mathcal{D}_-, \mathcal{D}_+)$ is a cyclic LP-structure with

$$\mathcal{D}_+ = V\{f \in C^\infty \cap \hbar_E : f'_1(s) + f_2(s) = 0, \, s > 0, \, f_2(0) = 0\},$$

$$\mathcal{D}_- = V\{f \in C^\infty \cap \hbar_E : f'_1(s) - f_2(s) = 0, \, s > 0, \, f_2(0) = 0\},$$

$$\mathcal{K} = (\mathcal{D}_- \oplus \mathcal{D}_+)^\perp.$$

9.8. Remarks

1. Ford, Kac, and Mazur [5] also discussed quantum systems. We recover their quantum process by replacing the Wiener mapping by a Weyl mapping associated with a quasi-free Kubo–Martin–Swchwinger state of the CCR algebra (see [19] and [12]).

2. We have confined our attention to stationary solutions of the Langevin equation. Non-stationary solutions will be treated in [11] together with a discussion of models which approximate to the Ford–Kac–Mazur model in the weak-coupling limit.

Acknowledgements

One of us (J.T.L.) would like to thank Professor M. Kac for introducing him to the Ford–Kac–Mazur model, and Dr. E. B. Davies, Dr. A. J. O'Connor, Dr. J. V. Pulè, and Dr. K. Schmidt for many discussions of the material.

References

1. DOOB, J. L. *Ann. Math.* **43**, 351 (1942).
2. DOOB, J. L. *Trans. Am. math. Soc.* **42**, 107 (1937).
3. EINSTEIN, A. *Annln Phys.* **17**, 549 (1905).
4. EINSTEIN, A. *Annln Phys.* **19**, 371 (1906).
5. FORD, G. W., KAC, M., and MAZUR, P. *J. Math. Phys.* **6**, 505–15 (1965).
6. KAC, M. *Lectures in differential equations*, Vol. 2 (ed. A. K. Aziz). Van Nostrand, New York (1969).
7. KIM, S. *J. Math. Phys.* **15**, 578 (1974).
8. LAMB, H. *Proc. Lond. math. Soc.* **2**, 88 (1900).
9. LANGEVIN, P. *C. r. hebd. Séanc. Acad. Sci., Paris* **146**, 530 (1908).
10. LAX, P. D. and PHILLIPS, R. S. *Scattering theory.* Academic Press, New York (1967).
11. LEWIS, J. T. and PULE, J. V. (In course of preparation.)
12. LEWIS, J. T. and THOMAS, L. C. Submitted to *Ann. Inst. H. Poincaré.*
13. LEWIS, J. T. and THOMAS, L. C. Z. *Wahrscheinlichkeitstheorie.* (To appear.)
14. NELSON, E. *Dynamical theories of brownian motion.* Princeton University Press (1967).
15. PERRIN, J. *C. r. hebd. Séanc. Acad. Sci., Paris* **149**, 477 (1909).
16. VON SMOLUCHOWSKI, M. *Annln Phys.* **21**, 756 (1906).

17. Sz-NAGY, B. *Acta Sci. math. Szeged* **15**, 87 (1953).
18. Sz-NAGY, B. and FOLIAS, C. *Harmonic analysis of operators on Hilbert space.* North-Holland, Amsterdam (1970).
19. THOMAS, L. C. D. Phil. Thesis, Oxford (1971).
20. UHLENBECK, G. E. and ORNSTEIN, L. S. *Phys. Rev.* **36**, 823 (1930).
21. WANG, M. C. and UHLENBECK, G. E. *Rev. mod. Phys.* **17**, 323 (1945).

10. Functional integrals and local many-body problems: localized moments and small particles

B. MÜHLSCHLEGEL

IN this chapter we discuss functional integrals in statistical physics that are introduced by eliminating the interaction between particles. First, some of the methods used for extended homogeneous many-body systems are considered. Local systems need different approximations due to larger fluctuations. This is illustrated on an atomic scale for the Anderson model of localized magnetic moments, and—on a somewhat larger scale—for superconductivity in minute particles.

10.1. Method of functional integrals for general many-body systems

During the 1950s, Green's functions and Feynman diagrams became widely used for sudying the finite-temperature properties of many-particle systems. In the wake of these activities functional averages were also introduced to statistical physics; some of the early applications have already been discussed, 11 years ago, by Edwards and by Siegert [1]. In this section we shall collect together the main ideas of functional integrals applied to equilibrium statistics of many particles, before considering more recent developments, in the subsequent two sections.

We restrict ourselves to homogeneous systems with direct two-body interactions. Then the Hamiltonian H (including $-\mu N$, where μ is the chemical potential), besides the part H_0 describing free particles, contains the interaction H_1, which can always be written as a sum of terms each of the form $A \cdot B$, where A and B are one-body operators which commute with each other,

$$H_1 = \sum_i A_i \cdot B_i, \qquad [A_i, B_i] = 0. \tag{10.1}$$

The statistical operator is represented as a functional integral (a Gaussian functional average) over a 'time'-ordered exponential

$$\exp(-\beta H) \equiv \exp[-\beta(H_0 + AB)] \tag{10.2}$$

$$= \int \delta z(\tau) \exp\left\{\frac{\pi}{\beta} \int_0^\beta d\tau |z(\tau)|^2\right\} \times$$

$$\times \exp\left\{-\int_0^\beta d\tau \left[H_0 + \sqrt{\left(\frac{\pi}{\beta}\right)} A z(\tau) - \sqrt{\left(\frac{\pi}{\beta}\right)} B z^*(\tau)\right]\right\}.$$

Here $\beta = (1/k_B T)$, $\delta z = \delta \operatorname{Re} z \cdot \delta \operatorname{Im} z$, and the summation in (10.1) has been dropped for the sake of simplicity. Feynman time-ordering, and in turn the functional integral, are necessary when H_0 does not commute with A and B. Eqn (10.2) follows by applying to the squares in

$$A \cdot B = -\tfrac{1}{4}(A-B)^2 + \tfrac{1}{4}(A+B)^2 \tag{10.3}$$

the identity

$$\exp(\pi q^2) = \int_{-\infty}^{\infty} ds \, \exp(-\pi s^2 - 2\pi q s) \tag{10.4}$$

so that $A - B$ is coupled with $\operatorname{Re} z$, and $A + B$ with $\operatorname{Im} z$, respectively.

The representation (10.2) is sometimes called the 'Stratonovich–Hubbard trick' since it was taken over from field theory to quantum statistics by Stratonovich [2] and in 1959 enlarged by Hubbard [3]. We should mention, however, that there are earlier applications in classical statistics. Edwards, amusingly enough, traces back the procedure even to Rayleigh (cf. [1]).

The partition function $Z = \operatorname{tr} \exp(-\beta H)$ becomes with eqn (10.2),

$$Z = \int dz \, e^{-\beta F(z,\beta)}, \tag{10.5a}$$

$$\beta F(z, \beta) = \frac{\pi}{\beta} \int_0^{\beta} d\tau |z(\tau)|^2 - \ln \operatorname{tr} \exp\left\{ -\int_0^{\beta} d\tau \left[H_0 + \sqrt{\frac{\pi}{\beta}} A z(\tau) - \sqrt{\frac{\pi}{\beta}} B z^*(\tau) \right] \right\}. \tag{10.5b}$$

This looks like a classical configuration integral with the coordinates replaced by the function $z(\tau)$. Actually, the second part of $F(z, \beta)$ is the thermodynamic potential of free particles (since A and B together with H_0 are one-body operators) moving in a τ-dependent potential. The Gaussian average over this potential then links, so to speak, infinitely many free-particle systems with the one interacting many-particle problem under consideration. Whether or not this bears any advantage for practical applications depends on the special system, and to a certain extent also on the affection of the researcher for this kind of mathematics. It should be said that all results obtained so far for conventional systems have also been obtained by means of diagrams, although some approximations look especially simple in the frame of functional averages, and in fact were introduced because of this simplicity.

The most obvious approximation is to look for a stationary point z_o of $F(z, \beta)$ and to expand F at z_0; in shorthand notation,

$$F = F(z_0) + \tfrac{1}{2} F''(z_0)(Z - z_0)^2 + O[z - z_0)^3]. \tag{10.6}$$

In case of normal many-body systems where one uses the decomposition of eqn. (10.1) in momentum-space,

$$H_1 = \frac{1}{2\Omega} \sum_q v(q) \rho_q^+ \rho_q, \qquad (10.7)$$

with $v(q)$ being the Fourier transform of the two-body interaction and Ω the volume of the system, z_0 and $F(z_0)$ correspond to the density ρ and thermodynamic potential, respectively, in Hartree approximation, whereas the evaluation of (10.5) with the quadratic term of (10.6) at once yields the random-phase (RPA) contributions to $\ln Z$.

Another quite intuitive approximation is based upon a variational procedure similar to the one used by Feynman [4] for the polaron path-integral. Choose a trial functional F_t and replace in (10.5a),

$$\exp(-\beta F) \approx \exp(-\beta F_t)[1 - \beta(F - F_t)]. \qquad (10.8)$$

The only workable *Ansatz* for F_t, again in shorthand notation, is

$$F_t = \alpha(z-a)^2, \qquad (10.9)$$

where a and α have to be determined by the stationarity of $\ln Z(a, \alpha)$. This procedure was first applied to the Ising model [5], in which case there is, of course, no time ordering necessary in eqn (10.2), since all parts of H commute; rather, the infinite-dimensional Gaussian integral comes about from H_1 in eqn (10.1) being the Ising exchange energy, the number of terms going to infinity in the thermodynamic limit. The variational parameters are very physical, since a corresponds to the magnetization and α is related to the spin-correlation function. The results of this calculation can be compared with reciprocal-range expansions, as was, for instance, discussed by Siegert [1]. We have here a nice analytical picture of the phase transition, although the results about the critical behaviour have been made rather obsolete by recent techniques (scaling, Wilson theory).

The variational procedure for normal many-body systems with an interaction given by eqn (10.7) was studied in extenso by Zittartz [5]. Here a and α correspond to the density and the two-particle correlation function respectively. The thermodynamic potential is stationary with respect to both physical quantities, and compared to the above Hartree–RPA approximation, this renormalized RPA contains many more diagrams, namely, all those which are necessary to determine the generalized dielectric function self-consistently. We should notice that a quite similar treatment was performed by Everts [5] for the Gorkov interaction of superconductivity. However, since we shall deal with some aspects of superconductivity in § 10.3, we shall not go into detail here.

The remarks made so far certainly do not cover all efforts in the application of functional averages to extended many-particle systems. However, the treatment of the van der Waals gas [6] must be mentioned. Sometimes the functional average representation on its own is useful in proving general relations; an example of this is the elementary derivation of the Griffiths–Kelly–Sherman inequalities by Monroe and Siegert [7].

10.2. The localized magnetic moment problem

Let us consider the most simple model of an atom with respect to its magnetic properties. There is only one orbital with energy ε_d, the orbital may be occupied by two electrons of opposite spin which then feel a Coulomb repulsion U. The corresponding Hamiltonian is

$$H_d = \varepsilon_d(n_\uparrow + n_\downarrow) + U n_\uparrow n_\downarrow, \tag{10.10}$$

where the subscript d indicates that H_d idealizes a transition metal atom with incomplete d-shell. Here $n_s = d_s^+ d_s$, d_s^+ creates an electron of spin $s = \uparrow, \downarrow$ in the orbital. Clearly the choice of the energy parameters offers the possibility of either a magnetic or unmagnetic atom (ion). For $\varepsilon_d < 0 < \varepsilon_d + U$ the ground state of H_d is magnetic (and two-fold degenerate) since only one electron is present.

Now imagine that the above atom is planted as an impurity into a metal described by its extended conduction-band states,

$$H_{\text{band}} = \sum_{ks} \varepsilon_k c_{ks}^+ c_{ks}, \tag{10.11}$$

and that a matrix element V_k allows for hopping of electrons from the extended states to the localized atomic orbital and vice versa,

$$H_{\text{And}} = H_d + \sum_{ks} (V_k c_{ks}^+ d_s + V_k^* d_s^+ c_{ks}) + H_{\text{band}}. \tag{10.12}$$

This model was invented by Anderson [8] in 1961 in order to understand how magnetic ions are influenced by their environment in a pure host metal. Anderson treated his model Hamiltonian in Hartree approximation, which appears somewhat surprising at first glance. Almost everyone who sees the Anderson model for the first time thinks that an exact solution should be easily accessible. After all (10.12) is one-body physics, except for the modest $U n_\uparrow n_\downarrow$ term, which should be mastered almost as easily as in the trivial case (10.10) of the isolated impurity. This hope has not been fulfilled during the years; instead it became clear that H_{And} describes a truly fundamental model, of great importance not only for the basic understanding of magnetism but perhaps for other areas of physics as well [9].

Turning to functional averages [10] we recognize at once that only the 'modest' two-body term $Un_\uparrow n_\downarrow$ has to be attacked by this method. Since H_d does not commute with the hopping interaction

$$W = \sum_{ks}(V_k c^+_{ks}d_s + V^*_k d^+_s c_{ks}) \tag{10.13}$$

in (10.12), we have to use Feynman time-ordering and obtain in complete analogy to (10.5)

$$Z = \int \delta z\, e^{-\beta F} = \int \delta z\, \exp\left[-\frac{\pi}{\beta}\int_0^\beta d\tau |z(\tau)|^2\right] \operatorname{tr} \exp\left\{-\int_0^\beta d\tau[E_\uparrow(\tau)n_\uparrow + E_\downarrow(\tau)n_\downarrow + W + H_{\text{band}}]\right\} \tag{10.14}$$

with

$$E_\uparrow(\tau) = \varepsilon_d - \sqrt{\left(\frac{\pi U}{\beta}\right)}z(\tau), \qquad E_\downarrow = \varepsilon_d + \sqrt{\left(\frac{\pi U}{\beta}\right)}z^*(\tau). \tag{10.15}$$

Under the integral in (10.14) the localized spin-up and spin-down d-electrons move independently in a time-dependent potential determined by

$$z(\tau) = x(\tau) + iy(\tau), \tag{10.16}$$

where, according to (10.15), $x(\tau)$ corresponds to a magnetic field and $iy(\tau)$ to an electric potential. The band states given by H_{band} only serve to broaden the d-electron levels via the hopping interaction W. In a broad-band approximation the d-level width becomes

$$\Gamma = \pi|\bar{V}|^2 N(0), \tag{10.17}$$

$\bar{V}$ and $N(0)$, being the hopping-matrix element and the density of band states, respectively, at the Fermi energy. Once this is accounted for, the extensive band part in (10.14) can be forgotten, and Z represents a truly local many-body problem.

After what has been said in § 10.1 it comes as no surprise to us that the stationary point z_0 of F in (10.14) just reproduces Anderson's original self-consistent Hartree solution. Since this solution contains a sharp transition from the magnetic to the nonmagnetic state with decrease of the relevant parameter

$$\eta = U/\pi\Gamma, \tag{10.18}$$

it is, in addition, not too surprising that an inclusion of RPA-contributions of the type indicated in eqn (10.6) will lead to an instability, characteristic to a phase transition. Since such a sharp phase transition is artificial for a local system where no thermodynamic limit is involved, RPA offers no improvement compared to the initial Hartree approximation. The same is to be expected for the more elaborate variation procedure of eqns (10.8) and (10.9) since it renormalizes RPA, and by this the phase transition is not softened but simply shifted away slightly. Obviously, the small fluctuation assumption made in (10.6) and (10.9), though useful for infinite systems, is no good for the local problem considered here.

The functional-average representation (10.14) suggests another approximation: simply drop time-ordering and consider (10.14) as a normal two-dimensional integral. This will be called the 'static approximation'. By means of Fourier transformation

$$z(\tau) = \frac{1}{\beta} \sum_{\nu=-\infty}^{\infty} z_\nu \, e^{-2\pi i \nu \tau/\beta} \tag{10.19}$$

(10.14) is transformed into an infinite-dimensional integral. The static approximation then neglects all dynamical fluctuations z_ν, $\nu \neq 0$. The remaining integral is still non-trivial. It is, however, interesting that the two Gaussian integrations coming from $z = x + iy$ can be carried out by the use of Euler's beta function, and we obtain the representation [11]

$$Z_{\text{stat}} = Z_{\text{band}} \frac{(2\pi)^2}{\left[\left(\frac{\beta\Gamma}{\pi} - 1\right)!\right]^2} \int_0^1 dn_\uparrow \, dn_\downarrow \, e^{-\beta f(n_\uparrow, n_\downarrow)} \tag{10.20a}$$

with

$$f(n_\uparrow, n_\downarrow) = \varepsilon_d(n_\uparrow + n_\downarrow) + U n_\uparrow n_\downarrow - (\Gamma/\pi - 1/\beta) \ln[4 \sin(\pi n_\uparrow) \sin(\pi n_\downarrow)]. \tag{10.20b}$$

Note that $n_\uparrow$, $n_\downarrow$ in (10.20) are continuous variables between 0 and 1, whereas before the same symbols were used for the d-electron number operators. Anderson's Hartree solution follows from $\partial f/\partial n_\uparrow = \partial f/\partial n_\downarrow = 0$ and contains his magnetic state (two minima of f with $\bar{n}_\uparrow \neq \bar{n}_\downarrow$) for the proper choice of parameters. Clearly, there is no justification for the symmetry-breaking saddle-point approximation of (10.20a). The static approximation preserves the ↑↓-symmetry. Moreover, it becomes exact in the limiting cases $\eta = U/\pi\Gamma \to 0, \infty$ and gives the right first-order corrections to these limits. How good is it then for finite η? The straight answer to this question must be delayed until someone comes along who can really calculate a functional integral of the type given by eqn (10.14). In the meantime, we should look for further approximations, and it is encouraging to see that the functional representation seems to offer novel approximations.

As an example refer to the recent work of Amit and Keiter [11], who go beyond the static approximation by including the components $z_{\nu = \pm 1}$ in addition to $z_{\nu=0}$ in eqn (10.19). They argue that this 'harmonic approximation' already contains essential dynamic corrections (unfortunately not the Kondo effect for $\eta \gg 1$). Amit and Keiter have numerically calculated the temperature-dependence of the magnetic susceptibility for different values of η. The result is shown in Fig. 10.1. As is to be expected, there is a *smooth transition from a Curie law to a temperature-independent Pauli susceptibility with decreasing η* both in static and harmonic approximation. The influence of dynamical fluctuations is strongest for $\eta \approx 1$ and leads to a reduction of the effective magnetic moment.

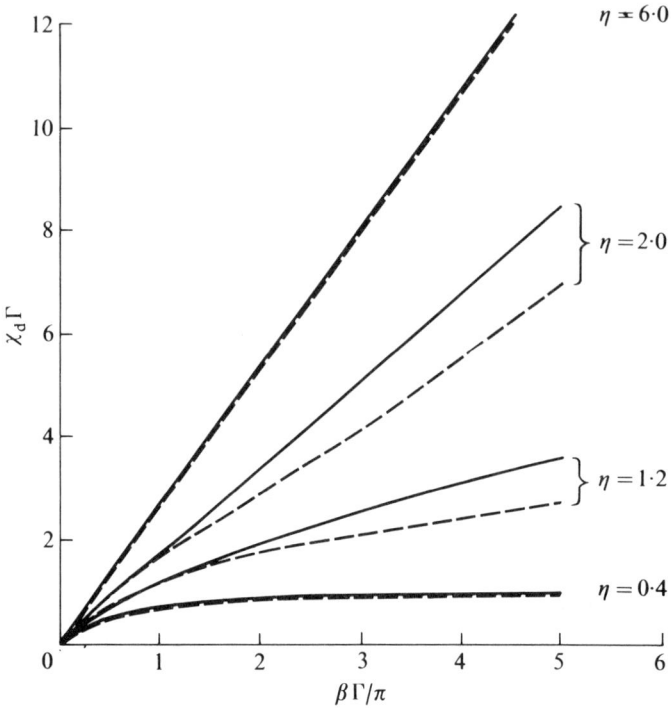

FIG. 10.1. Excess susceptibility due to the impurity versus $\beta\Gamma/\pi$ for different values of $\eta = U/\pi\Gamma$. Solid curves represent the static approximation, dashed curves are for the harmonic approximation. (Figure taken from ref. [11].)

So far we have discussed a special extension of the static approximation. To overcome the failure of RPA, Schrieffer and co-workers have extended the Taylor expansion of eqn (10.6) by including quartic terms. This work is covered in a recent review by Hamann and Schrieffer [12], and will not be described here. A very interesting problem is connected with the fact that the model of an impurity spin coupled antiferromagnetically to the band electrons (the famous Kondo model) is contained in the magnetic Anderson model and should even take over its essential physical properties for $\eta \gg 1$

[13]. We cannot discuss the details of the Kondo effect here, but it is necessary to mention, in this context, the work of Hamann. By a sequence of approximations he was able to transform the functional integral given by eqn (10.14) into

$$Z = Z_{V=0} \int \delta x \, \exp\left[-\int_0^\beta d\tau V(x) - \frac{1}{\beta}\int_0^\beta d\tau \, d\tau_1 \, T(x, x_1, x_1', \tau, \tau_1)\right], \quad (10.21)$$

which resembles a one-dimensional Feynman path-integral. Hamann then selects paths $x(\tau)$ which correspond to a hopping back and forth between the two minima of Anderson's magnetic state in Hartree approximation. This is not yet the final solution but it brings the problem in close contact with approximations performed for the Kondo model. For details and further references consult the above-mentioned review article by Hamann and Schrieffer [12].

10.3. Zero-dimensional superconductors

Two-body interactions responsible for superconductivity of an electronic system can be treated by means of functional integrals as was pointed out in § 10.1. In a highly simplified version the Bardeen–Cooper–Schrieffer (BCS) interaction [14] is

$$H_1 = -\frac{V}{\Omega} B^+ B, \quad (10.22a)$$

with

$$B^+ = \sum_k b_k^+, \qquad b_k^+ = c_{k\uparrow}^+ c_{-k\downarrow}^+. \quad (10.22b)$$

As was the case in eqn (10.11) $c_{k\uparrow}^+$ creates an electron of momentum k and spin-up in a conduction-band state. b_k^+ creates a pair of electrons with opposite momenta and spins. $V > 0$ is the effective interaction and Ω the volume. The Hamiltonian H_0 of free particles is identical with H_{band} of eqn (10.11). In the large-volume limit $H_0 + H_1$ describes an extended system, and the operators B^+ and B practically commute with each other. Therefore we can immediately use the formulas (10.2) and (10.5) of § 10.1 with the proper identification of the operators A and B. The resulting BCS functional integral has been studied in detail for large Ω [15]. Due to the symmetry of (10.22a) the minimum of $F(z, \beta)$ is reached not at an isolated point z_0 but at a circle $|z_0| = r_o(\beta)$. The corresponding Hartree-like thermodynamic potential

$$F_{\text{BCS}}(\beta) = F(z_0(\beta), \beta) \quad (10.23)$$

gives us the well-known microscopic BCS-description of the superconducting phase, with z_0 being directly proportional to the famous temperature-dependent energy gap, or order parameter.

132   *Functional integrals and local many-body problems:*

When we take into account static and dynamic fluctuations in the quadratic approximation as indicated by eqn (10.6) and calculate their contributions to ln Z we find that these have a lower volume-dependence than F_{BCS}. Therefore the BCS thermodynamic potential becomes exact in the thermodynamic limit. The origin of this result lies in the special truncated form (10.22) of the chosen model. Similarly, the molecular field approximation of the Ising model becomes exact for the limiting case of infinite range and vanishing interaction.

Let us finally turn to the subject indicated by the somewhat provoking title of this section. What happens to superconductivity when we do not consider bulk systems but when, instead, Ω becomes smaller and smaller? As is well known, the BCS theory has a characteristic pair coherence length ξ built in, which may be of order 1000 Å or more, and which has to be considered as the natural length for the physical phenomena of interest. Because of this, systems of dimensions less than the coherence length (e.g. metal particles of order 100 Å in size) may be called local or zero-dimensional. What are then the properties of such local superconductors?

The framework of functional averages suggests here a simple approach to answer this question. Clearly, the steepest-descents expansion used for the bulk with $\xi^{-3}\Omega$ being the large-expansion parameter is no good for the small system. Instead we may approximate the BCS functional integral statically by dropping the τ-dependence in eqn (10.5) and converting Z in a normal two-dimensional integral Z_{stat}. The trace in (10.5b) can be evaluated; moreover, putting $z = r \exp(i\phi)$, the ϕ-integration (which takes care of particle-number conservation) can be performed and we obtain

$$Z_{stat} = Z_0 \frac{\beta}{g\delta} \int_0^\infty d|\psi|^2 \exp[-\beta f(|\psi|^2, \beta)] \quad (10.24a)$$

with

$$f(|\psi|^2, \beta) = \frac{|\psi|^2}{g\delta} - \frac{2}{\beta} \sum_\alpha^1 \ln \frac{\cosh(\beta E_\alpha/2)}{\cosh(\beta \varepsilon_\alpha/2)}, \qquad E_\alpha = \sum(\varepsilon_\alpha^2 + |\psi|^2). \quad (10.24b)$$

Here, ψ may be identified with the BCS order parameter. In the small system the one-electron states are labelled by α, The energy is ε_α. The pairing interacting (10.22) is between time-reversed states $\alpha, \tilde{\alpha}$, and we have put

$$\frac{V}{\Omega} = g\delta, \qquad \delta = \frac{1}{N(0)\Omega} \quad (10.25)$$

by introducing a dimensionless coupling constant g. The important parameter in (10.24) is δ, the reciprocal density of states of the whole system at the Fermi energy. Z_0, eventually, denotes the partition function of free electrons.

The expression (10.24) has recently been used to calculate various properties of small superconducting particles [16]. Since the sum in (10.24b) contains a characteristic cut-off of energies $|\varepsilon_\alpha| > \omega_D$, the transition temperature T_c can be introduced as usual by

$$k_B T_c = 1\cdot 14\, \omega_D\, e^{-1/g}. \tag{10.26}$$

Actually, T_c would be the transition temperature of a bulk system with unchanged coupling constant g. Since the energy $\delta = [N(0)\Omega]^{-1}$ is the average level-spacing of single electron states at the Fermi energy in the finite system, the most natural parameter to measure size effects is

$$\bar\delta = \delta / k_B T_c. \tag{10.27}$$

In Fig. 10.2 the specific heat (C) is shown as an example, and it is seen how the specific-heat anomaly of the bulk system is washed out as $\bar\delta$ increases. It

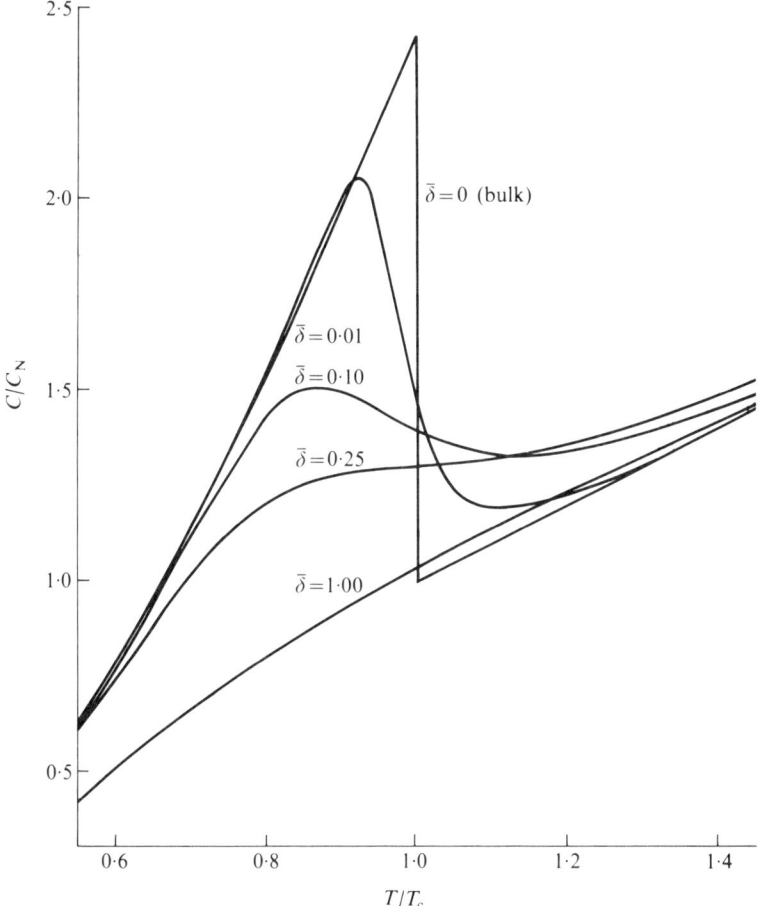

FIG. 10.2. Normalized specific heat calculated in the static functional approximation for several values of $\bar\delta = \delta/k_B T_c$. $\bar\delta = \delta/k_B T_c$. reference the bulk BCS limit is also shown. (Figure taken from ref. [16].)

is interesting that when $\bar{\delta}$ approaches unity there is hardly a remnant of the superconducting phase transition at $T = T_c$. Concerning the numerical calculations for finite $\bar{\delta}$ we have to realize that the size effect enters (10.24) in a two-fold manner. First, the sum is discrete in (10.24b), which is taken care of by an equal-level-spacing description; and second the integration has to be performed without saddle-point approximation in order to include all static fluctuations.

Of course, measurements are done, not on a single small particle but on probes which contain many of them, all isolated from each other and possibly all of the same size. The specific heat is a difficult quantity to measure, but there are other physical properties which are more accessible and which have confirmed the usefulness of our simple model. Size effects of a similar type should also be expected in small droplets of superfluid He^4 but, to the author's knowledge, have not yet been observed.

In conclusion we may ask: How good is the static approximation from a theoretical point of view? In superconductivity, spatial variations of the order parameter occur on a scale given by the coherence length ξ. Therefore it seems to be justified that we use a spatially constant order parameter for sufficiently small particles. It may be more serious that all dynamic fluctuations have been neglected. However, recent work by Hassing and Wilkins [17] shows that for the chosen model the dynamical fluctuations contained in the functional integral are dominated by the static ones, at least in the critical region near T_c. This then, fortunately, puts our static model on a somewhat sound basis.

Acknowledgements

It is a pleasure for me to thank Professor H. Keiter for some useful discussions.

References

1. *Proceedings of a Conference on theory and applications of analysis in function space, 1963* (eds. W. T. Martin and I. Segal). The M.I.T. Press, Cambridge, Mass. (1964). See S. F. Edwards (p. 31 and p. 167) and A. J. F. Siegert (p. 154). See these articles also for earlier literature.
2. STRATONOVICH, R. L. *Dokl. Akad. nauk. SSSR* **115**, 1097 (1957). Translated *Sov. Phys. Dokl.* **2**, 416 (1958).
3. HUBBARD, J. *Phys. Rev. Lett.* **3**, 77 (1959). A general and more recent contribution which also includes Green's functions and linear response is due to SHERINGTON, D. *J. Phys. C* **4**, 401 (1971).
4. FEYNMAN, R. P. and HIBBS, A. R. *Quantum mechanics and path integrals.* McGraw–Hill, New York (1965).
5. MÜHLSCHLEGEL, B. and ZITTARTZ, J. *Z. Phys.* **175**, 553 (1963); ZITTARTZ, J. *Z. Phys.* **180**, 219 (1964); EVERTS, U. *Z. Phys.* **199**, 211 (1967).
6. WIEGEL, F. W. *J. stat. Phys.* **7**, 213 (1973), and references given therein.
7. MONROE, J. L. and SIEGERT, A. J. F. To be published.
8. ANDERSON, P. W. *Phys. Rev.* **124**, 41 (1961).

9. e.g. adsorption at surfaces: SCHRIEFFER, J. R. *J. Vac. Sci. Technol.* **9**, 561 (1972).
10. MÜHLSCHLEGEL, B. Unpublished lecture notes. University of Pennsylvania (1965).
11. AMIT, D. J. and KEITER, H. *J. low Temp. Phys.* **11**, 603 (1973).
12. HAMANN, D. R. and SCHRIEFFER, J. R. In *Magnetism* (ed. H. Suhl), Vol. V, pp. 237–52. Academic Press, New York (1973).
13. This was first shown by a canonical transformation due to SCHRIEFFER, J. R. and WOLFF, P. R. *Phys. Rev.* **149**, 491 (1966), and by perturbative calculation of the susceptibility, SCALAPINO, D. J. *Phys. Rev. Lett.* **16**, 937 (1966).
14. BARDEEN, J., COOPER, L. N., and SCHRIEFFER, J. R. *Phys. Rev.* **108**, 1175 (1957).
15. MÜHLSCHLEGEL, B. *J. Math. Phys.* **3**, 522 (1962).
16. MÜHLSCHLEGEL, B., SCALAPINO, D. J., and DENTON, R. *Phys. Rev.* **B6**, 1767 (1972).
17. HASSING, R. F. and WILKINS, J. W. *Phys. Rev.* **B7**, 1890 (1973).

11. Remarks on Markov field equations

E. NELSON

11.1. Introduction

WE have no new existence theorems in constructive quantum field theory to present here, but we wish to indicate a new direction which looks promising and which certainly poses many interesting questions.

Only the theory of a neutral scalar field with a quartic self-interaction will be considered. This theory has been much studied in dimension $d = 2$ and Glimm and Jaffe have pioneered the study in dimension $d = 3$. (For references, see [4]—in particular, see the first article by Glimm, Jaffe, and Spencer and the reference listed there.)

We wish to stress field equations, and so we will begin with a heuristic discussion from that point of view. The simplest non-linear relativistic field equation with good formal properties is

$$(\Box + m^2)A = -gA^3 + \alpha A, \tag{11.1}$$

corresponding to the interaction Lagrangian density $-(g/4)A^4 + (\alpha/2)A^2$. We could of course absorb the linear term αA in the term $m^2 A$, but we prefer not to. Here m^2 and g are positive.

The Euclidean approach to the problem of quantized solutions to (11.1) is, in rough outline, as follows. The Wightman distributions (vacuum expectation values of products of field operators) are the boundary values of analytic functions which in the Euclidean region of points with real space and imaginary time coordinates are known as the Schwinger functions. The Schwinger functions are the expectation values of products of a random field ϕ. In the free case ($g = \alpha = 0$) the field ϕ is the Gaussian field of mean 0 and covariance G, where G is the kernel of the integral operator $(-\Delta + m^2)^{-1}$. Because of the singularity at the origin (for $d \geq 2$) ϕ is a generalized random field or random distribution (see [1]).

The Schwinger functions in the interacting case are formally given by changing the measure to the new measure with density

$$\frac{\exp\left\{\int\left[-\frac{g}{4}\phi(x)^4 + \frac{\alpha}{2}\phi(x)^2\right]dx\right\}}{E_0 \exp\left\{\int\left[-\frac{g}{4}\phi(x)^4 + \frac{\alpha}{2}\phi(x)^2\right]dx\right\}}, \tag{11.2}$$

where the integration is over d-dimensional Euclidean space $\mathbf{E}^d$ and E_0 denotes expectations with respect to the free Gaussian measure. An important property of the free Euclidean field is the Markov property, which asserts that knowledge of the field outside a region Λ gives no more information about the behaviour of the field inside Λ than does knowledge of the field on the boundary of Λ. This property is preserved under the change of measure due to the local nature of the integrand.

From a knowledge of the Schwinger functions we reconstruct the Wightman distributions and the quantum field A. Glimm and Jaffe [2] show that in dimension $d=2$ the field A satisfies (11.1) where, formally, α is the infinite constant $3gE_0\phi(x)^2$. It must be emphasized that the Euclidean field ϕ does not satisfy the equation $(-\Delta + m^2)\phi = 0$. If it did, then by the regularity theorem for elliptic partial differential equations, ϕ would be a real analytic function with probability one, which is far from being true. Let $(-\Delta + m^2)\phi = \omega$, where ϕ is the free Euclidean field. Then ω is a Gaussian random field with mean 0 and covariance $(-\Delta + m^2)\delta$. In the interacting case, we set $(-\Delta + m^2)\phi = -g\phi^3 + \alpha\phi + \omega$, and we shall begin the study of the term ω defined in this way. To make the discussion meaningful, especially the change of measure (11.2), it is helpful to introduce the lattice approximation.

11.2. The lattice approximation

Consider the lattice $\varepsilon \mathbf{Z}^d$. We give $\varepsilon \mathbf{Z}^d$ the measure which assigns weight ε^d to each point, so that if f is a function on $\varepsilon \mathbf{Z}^d$ then

$$\int f(x)\, dx = \sum f(x)\varepsilon^d.$$

We define

$$\delta(x-y) = \begin{cases} 1/\varepsilon^d, & x = y \\ 0, & x \neq y. \end{cases}$$

Fix a subset Λ of $\varepsilon \mathbf{Z}^d$. If f is a function defined on Λ we let

$$\Delta f(x) = \frac{1}{\varepsilon^2}\left[-2df(x) + \sum_{\substack{|y-x|=\varepsilon \\ y \in \Lambda}} f(y)\right], \qquad x \in \Lambda.$$

Thus Δ is the discrete analogue of the Laplace operator with Dirichlet boundary condition. Let G be the kernel of the operator $(-\Delta + m^2)^{-1}$, where m^2 is a positive constant (and is strictly positive if Λ is the entire lattice $\varepsilon \mathbf{Z}^d$ for $d=1$ or $d=2$). Thus

$$(-\Delta + m^2)^{-1} f(x) = \sum_{y \in \Lambda} G(x,y) f(y) \varepsilon^d, \qquad x \in \Lambda.$$

Then G is a function of positive type on $\Lambda \times \Lambda$. Let A be the kernel of $(-\Delta + m^2)$, so that

$$A(x, y) = \begin{cases} 2d\varepsilon^{d-2} + m^2 \varepsilon^d, & x = y \\ -\varepsilon^{d-2}, & |x-y| = \varepsilon \\ 0, & |x-y| > \varepsilon. \end{cases} \quad (11.3)$$

We let ϕ be the Gaussian process indexed by Λ with mean 0 and covariance G. If Λ is finite then the $\phi(x)$ are just the coordinate functions on $\mathbf{R}^{\#\Lambda}$ with respect to the Gaussian measure

$$\exp(-\frac{1}{2} \sum_{x,y \in \Lambda} \xi_x A(x, y) \xi_y)(2\pi)^{-\#\Lambda/2}(\det A)^{-\frac{1}{2}} \prod_{x \in \Lambda} d\xi_x. \quad (11.4)$$

Suppose that the values of $\phi(y)$ are known for the nearest neighbours y of x (i.e. for these points y in Λ for which $|x - y| = \varepsilon$). Then the conditional probability distribution of $\phi_0(x)$ is given by

$$\frac{1}{\sqrt{(2\pi\sigma^2)}} \exp(-(\xi - \mu)^2 / 2\sigma^2) \, d\xi,$$

where

$$\mu = \frac{1}{2d + m^2\varepsilon^2} \sum_{|y-x|=\varepsilon} \phi(y) \quad (11.5)$$

and

$$\sigma^2 = \frac{1}{2d\varepsilon^{d-2} + m^2\varepsilon^d}. \quad (11.6)$$

We use E_0 to denote expectations with respect to the measure (11.4).

Now let g be positive and α real and assume that Λ is finite. We give $\mathbf{R}^{\#\Lambda}$ the new probability measure whose density with respect to the Gaussian measure (11.4) is

$$\frac{\exp\left[\sum_{x \in \Lambda} \left(-\frac{g}{4}\xi_x^4 + \frac{\alpha}{2}\xi_x^2\right)\varepsilon^d\right]}{E_0 \exp\left[\sum_{x \in \Lambda} \left(-\frac{g}{4}\xi_x^4 + \frac{\alpha}{2}\xi_x^2\right)\varepsilon^d\right]}. \quad (11.7)$$

Again we let the $\phi(x)$ be the coordinate functions on $\mathbf{R}^{\#\Lambda}$, so that ϕ is a stochastic process indexed by Λ which is non-Gaussian if $g \neq 0$. We use E to denote expectations with respect to the new measure.

Suppose that the values of $\phi(y)$ are known for the nearest neighbours y of x. Then the conditional probability distribution of $\phi(x)$ is given by

$$\frac{1}{N} \exp\left[\left(-\frac{g}{4}\xi^4 + \frac{\alpha}{2}\xi^2\right)\varepsilon^d\right] \exp\left[\frac{-(\xi-\mu)^2}{2\sigma^2}\right] d\xi,$$

where μ and σ^2 are given as before by (11.5) and (11.6) respectively (notice, however, that μ is now a different random variable because the measure has been changed) and N is the normalization constant

$$N = \int_{-\infty}^{\infty} \exp\left[\left(-\frac{g}{4}\xi^4 + \frac{\alpha}{2}\xi^2\right)\varepsilon^d\right] \exp\left[-\frac{(\xi-\mu)^2}{2\sigma^2}\right] d\xi.$$

Notice that in the measure (11.4), and in the new measure with density (11.7), only nearest neighbours are coupled, so that the conditional expectation of any function of $\phi(x)$ with respect to the $\phi(y)$ for all $y \neq x$ is the same as its conditional expectation with respect to the $\phi(y)$ for the nearest neighbours y of x. This is the Markov property. We use $\hat{E}$ (or $\hat{E}_x$ if necessary to avoid ambiguity) to denote such conditional expectations.

11.3. The Markov field equation

Let us compute:

$$\hat{E}\phi(x) - \mu = \frac{1}{N} \int_{-\infty}^{\infty} (\xi - \mu) \exp\left[\left(-\frac{g}{4}\xi^4 + \frac{\alpha}{2}\xi^2\right)\varepsilon^d\right] \exp\left[-\frac{(\xi-\mu)^2}{2\sigma^2}\right] d\xi$$

$$= \frac{1}{N} \int_{-\infty}^{\infty} \exp\left[\left(-\frac{g}{4}\xi^4 + \frac{\alpha}{2}\xi^2\right)\varepsilon^d\right] \left(-\sigma^2 \frac{d}{d\xi} \exp\left[-\frac{(\xi-\mu)^2}{2\sigma^2}\right]\right) d\xi$$

$$= \frac{1}{N} \int_{-\infty}^{\infty} \left(\sigma^2 \frac{d}{d\xi} \exp\left[\left(-\frac{g}{4}\xi^4 + \frac{\alpha}{2}\xi^2\right)\varepsilon^d\right]\right) \exp\left[-\frac{(\xi-\mu)^2}{2\sigma^2}\right] d\xi$$

$$= \frac{1}{N} \int_{-\infty}^{\infty} \sigma^2(-g\xi^3 + \alpha\xi)\varepsilon^d \exp\left[\left(-\frac{g}{4}\xi^4 + \frac{\alpha}{2}\xi^2\right)\varepsilon^d\right] \times$$

$$\times \exp\left[-\frac{(\xi-\mu)^2}{2\sigma^2}\right] d\xi$$

$$= \sigma^2 \hat{E}[-g\phi(x)^3 + \alpha\phi(x)]\varepsilon^d.$$

We may write this as

$$\phi(x) - \mu = \sigma^2[-g\phi(x)^3 + \alpha\phi(x)]\varepsilon^d + \sigma^2\omega(x)\varepsilon^d, \quad (11.8)$$

where $\omega(x)$ is a function of $\phi(x)$ and $\phi(y)$ for the nearest neighbours y of x (because (11.8) is the definition of $\omega(x)$) and

$$\hat{E}\omega(x) = 0. \quad (11.9)$$

Using (11.5), (11.6), and the definition of Δ, we may rewrite (11.8) as

$$(-\Delta + m^2)\phi(x) = -g\phi(x)^3 + \alpha\phi(x) + \omega(x). \qquad (11.10)$$

We call this the *Markov field equation* (in the lattice approximation). We shall derive the basic properties of the stochastic term $\omega(x)$.

By (11.8),

$$\hat{E}\phi(x)^n \sigma^2 \omega(x)\varepsilon^d = \hat{E}\phi(x)^n\{\phi(x) - \mu - \sigma^2[-g\phi(x)^3 + \alpha\phi(x)]\varepsilon^d\}$$

$$= \frac{1}{N}\int_{-\infty}^{\infty} \xi^n[\xi - \mu - \sigma^2(-g\xi^3 + \alpha\xi)\varepsilon^d]\exp\left[\left(-\frac{g}{4}\xi^4 + \frac{\alpha}{2}\xi^2\right)\varepsilon^d\right] \times$$

$$\times \exp\left[-\frac{(\xi-\mu)^2}{2\sigma^2}\right]d\xi.$$

As before, we represent the $(\xi - \mu)$ term as $-\sigma^2(d/d\xi)$ acting on the Gaussian factor and then integrate by parts. The cancellation which occurs leaves us with

$$\frac{1}{N}\int_{-\infty}^{\infty}\sigma^2 n\xi^{n-1}\exp\left[\left(-\frac{g}{4}\xi^4 + \frac{\alpha}{2}\xi^2\right)\varepsilon^d\right]\exp\left[-\frac{(\xi-\mu)^2}{2\sigma^2}\right]d\xi = \hat{E}\sigma^2 n\phi(x)^{n-1}.$$

Therefore

$$\hat{E}\phi(x)^n \omega(x) = \hat{E}n\phi(x)^{n-1}\frac{1}{\varepsilon^d}.$$

If $y \neq x$ we have $\hat{E}\phi(y)^n \omega(x) = 0$, by the Markov property, so that if P is any polynomial then

$$\hat{E}P(\phi(y))\omega(x) = \hat{E}P'(\phi(y))\delta(x-y). \qquad (11.11)$$

(In fact, the argument holds for much more general functions P.)

Let S be the Schwinger function

$$S(x_1, \ldots, x_n) = E\phi(x_1)\ldots\phi(x_n).$$

Then by (11.10) and (11.11) we have, using Δ_1 to denote the operator Δ with respect to the variable x_1,

$$(-\Delta_1 + m^2)S(x_1, x_2, \ldots, x_n) = -gS(x_1, x_1, x_1, x_2, \ldots, x_n) +$$
$$+ \alpha S(x_1, x_2, \ldots, x_n) + \sum_{i=2}^n \delta(x_1 - x_i)S(x_1, x_2, \ldots, \hat{x}_i, \ldots, x_n),$$
$$(11.12)$$

where $\hat{x}_i$ means to omit x_i (see [3]).

Finally, we compute some moments of the ω process.

We have

$$\hat{E}\omega(x)^2 = \hat{E}\frac{1}{\sigma^2\varepsilon^d}\{\phi(x)-\mu-\sigma^2[-g\phi(x)^3+\alpha\phi(x)]\varepsilon^d\}\omega(x) \tag{11.13}$$

$$= \hat{E}\frac{1}{\sigma^2\varepsilon^d}\left\{\frac{1}{\varepsilon^d}-\sigma^2[-3g\phi(x)^2+\alpha]\right\}$$

$$= \frac{1}{\sigma^2\varepsilon^{2d}}+\hat{E}[3g\phi(x)^2-\alpha]\frac{1}{\varepsilon^d}.$$

If x and y are nearest neighbours then

$$\hat{E}_y\omega(x)\omega(y) = \hat{E}_y\frac{1}{\sigma^2\varepsilon^d}\{\phi(x)-\mu-\sigma^2[-g\phi(x)^3+\alpha\phi(x)]\varepsilon^d\}\omega(y)$$

$$= \hat{E}_y\frac{1}{\sigma^2\varepsilon^d}(-\mu)\omega(y).$$

Using (11.5), (11.6), and $\hat{E}_y\phi(y)\omega(y)=1/\varepsilon^d$, we find that

$$\hat{E}_y\omega(x)\omega(y) = -\frac{1}{\varepsilon^{d+2}}. \tag{11.14}$$

Let

$$\tilde{m}^2 = m^2 + E[3g\phi(x)^2-\alpha]. \tag{11.15}$$

By (11.13), (11.14), and (11.4) we have

$$E\omega(x)\omega(y) = (-\Delta+\tilde{m}^2)\delta(x-y) \tag{11.16}$$

(since $E\omega(x)\omega(y)=0$ if $|x-y|>\varepsilon$).

If we compute the fourth moment of $\omega(x)$ we find

$$E\omega(x)^4 = 3[E\omega(x)^2]^2 + 3(E[-3g\phi(x)^2+\alpha]^2-\{E[-3g\phi(x)+\alpha]\}^2)\frac{1}{\varepsilon^{2d}}$$

$$-18g\frac{1}{\varepsilon^{3d}}, \tag{11.17}$$

so that ω is in general non-Gaussian.

11.4. Speculations

The hope of the Euclidean approach is to choose the counter term α depending on Λ in such a way that in the limit as Λ increases to the entire lattice $\varepsilon\mathbf{Z}^d$ and the lattice spacing ε decreases to 0, the Schwinger functions will have limits which are the Schwinger functions of a quantum field theory satisfying all of the Wightman axioms—except that the uniqueness of the vacuum (cluster decomposition property of the expectation values) may hold only in certain ranges of the parameters. In dimension $d=2$ this may be achieved by choosing $\alpha = 3gE_0\phi(x)^2$.

Suppose that this has been done. What can be said about the Markov field equation (11.10)? We may define the stochastic term ω as a quadratic form by

$$\langle \phi(x_1) \ldots \phi(x_n), \omega(x)\phi(x_{r+1}) \ldots \phi(x_n)\rangle = \sum_{i=1}^{n} \delta(x-x_i) S(x_1, \ldots, \hat{x}_i, \ldots, x_n),$$

so that $\int \omega(x) f(x) \, dx$ is a well-defined quadratic form if f is a smooth test-function (if assume that the singularities of the Schwinger functions at coinciding arguments are not too great). The left-hand side of the equation is well defined as a random variable.

$$\int [(-\Delta + m^2)\phi(x)] f(x) \, dx = \int \phi(x)[(-\Delta + m^2)f](x) \, dx.$$

Consequently, we may define $\int [-g\phi(x)^3 + \alpha\phi(x)] f(x) \, dx$ as a quadratic form by taking the difference of the other two terms.

Consider the formal expression $-g\phi(x)^3 + \alpha\phi(x)$ as a functional of the field. If we add a smooth numerical function Φ to ϕ we obtain

$$-g[\phi(x) + \Phi(x)]^3 + \alpha[\phi(x) + \Phi(x)]$$
$$= -g\phi(x)^3 + \alpha\phi(x) + [-3g\phi(x)^2 + \alpha]\Phi(x) + [-3g\phi(x)]\Phi(x)^2 - g\Phi(x)^3.$$

In order for this to have a well-defined expectation, it is necessary and sufficient that $E[-3g\phi(x)^2 + \alpha]$ be finite. This is the problem of 'Wick ordering with respect to the physical vacuum', and it would be nice if it could be achieved. It would also be nice if $-g\phi(x)^3 + \alpha\phi(x)$, smeared with a smooth test-function, were a well-defined random variable. A necessary and sufficient condition for this, by the Markov field equation, is that $\omega(x)$, smeared with a smooth test-function, be a well-defined random variable. But the covariance of ω is $(-\Delta + \tilde{m}^2)\delta$ which is well defined iff $\tilde{m}^2$ is finite. By (11.15), this is the case iff $E[-3g\phi(x)^2 + \alpha]$ is finite. Thus heuristically, the two desiderata are equivalent. In dimension $d=3$ it seems unlikely that they can be achieved. In fact, even for the free field we cannot interpret $-g\phi(x)^3 + \alpha\phi(x)$ as a random variable (except for the circumstance that g and α vanish!), since $:\phi(x)^3:$ has the covariance $(1/4\pi r)e^{-mr^3}$, where $r = |x-y|$, which has a logarithmically divergent integral.

References

1. GEL'FAND, I. M. and VILENKIN, N. YA. *Generalized functions*, Vol. 4, *Applications of harmonic analysis* (translated by A. Feinstein). Academic Press, New York (1964).

2. GLIMM, J. and JAFFE, A. The $\lambda(\phi^4)_2$ quantum field theory without cutoffs II, The field operators and the approximate vacuum. *Ann. Math.* **91**, 362–401 (1970).
3. SYMANZIK, K. Euclidean quantum field theory. In *Local quantum theory, Proceedings of the International School of Physics 'Enrico Fermi'*, Course 45 ed. R. Jost. Academic Press, New York (1969).
4. *Constructive quantum field theory, The 1973 'Ettore Majorana' International School of Mathematical Physics* (eds. G. Velo and A. Wightman). *Lecture notes in physics*, Vol. 25. Springer–Verlag, Berlin (1973).

12. Caustics and multivaluedness: two results of adding path amplitudes

L. S. Schulman

THE practitioner of functional integration whose main interest lies in quantum mechanics is of necessity a person with a guilty conscience. Nature has given us complex amplitudes, but mathematicians only allow us to add paths with real measures. Some of us contend with this burden of guilt by escape: escape to dual spaces and continuation to the very boundaries of good behaviour. Others, perhaps of sterner stuff, attempt to work in risky intimate contact with sin, though endeavouring at all times to keep their own hands clean. Those in the latter class talk and think in terms of *bona fide* sum-over-paths—complex amplitudes and all—but will also attempt to put their thoughts in the form of an iteration of integrals so as not to be led astray by intuition. But the process is risky, and there have been many casualties.

In this chapter, we will discuss two rather different kinds of work which have in common that they are easily stated and understood in 'global' terms, i.e. in the slippery language of the sum-over-paths. In the first of these applications we shall find that path-integrals are a good way to draw out the physical consequences that homotopy theory may have for a system—the importance of the conceptual summing over paths will be evident here. The second application goes back to the way in which the correspondence limit ($\hbar \to 0$) is made to appear 'obvious' in the path-integral formulation of quantum mechanics. Namely, since the propagator is a sum $K = \sum \exp[iS(x)/\hbar]$ then by the naïve application of the method of stationary phase, K will be dominated by the classical paths (i.e. those for which $\delta S = 0$) in the limit $\hbar \to 0$. We shall examine this limit with perhaps a bit less naïveté and will also see that some other developments concerned with uniform asymptotic expansions carry through to the path-integral—instead of coalescing extremum points we shall be concerned with the coalescing of classical paths.

12.1. Propagators for multivalued wave functions

First we will discuss the way in which quantum mechanics is enriched when one considers the path-integral on a multiply connected space. Let $K(b, a; T)$ be the propagator on a space M from a at time 0 to b at time T for some given Hamiltonian. If we take $b = a$ (which we do to avoid having to give a lot of definitions) then K can be written

$$K(a, a; T) = \sum_{g \in \pi} e^{i\nu(g)} \sum_{\gamma \in g} e^{iS(\gamma)/\hbar}, \qquad (12.1)$$

where π is the fundamental homotopy group of M, $g \in \pi$ is an equivalence class of paths, $\gamma \in g$ is a path beginning and ending at a, and the real numbers $\{\nu(g)\}$ form an additive representation of π,

$$\nu(g_1) + \nu(g_2) = \nu(g_1 g_2). \tag{12.2}$$

This result is used in references [1], [2], [3], and [4], and a detailed proof is given in [4]. Rather than go carefully into that proof the author will give a simple demonstration of how the form of the propagator in eqn (12.1) is in a sense unavoidable and how eqn (12.2) arises in a natural way.

Suppose we did not allow for the phase factors ν, and simply wrote the propagator

$$K(b, a; T) = \sum_\gamma \exp[iS(\gamma)/\hbar]$$

with

$$S(\gamma) = \int_0^T L \, dt = \int_0^T \left(\frac{1}{2}m v^2 + \frac{e}{c} \mathbf{v} \cdot \mathbf{A} - V\right) dt,$$

the integrals being evaluated along γ. Consider a gauge transformation $\mathbf{A} \to \mathbf{A} = \mathbf{A}' + \nabla \phi$. Then

$$K'(b, a; T) = \sum_\gamma \exp\left[\frac{iS(\gamma)}{\hbar} + \frac{ie}{\hbar c} \int_\gamma \nabla\phi \cdot d\mathbf{x}\right]. \tag{12.3}$$

Now there are two good reasons why it must be the case that

$$\int_\gamma \nabla\phi \cdot d\mathbf{x} = \phi(b) - \phi(a). \tag{12.4}$$

Before going into these reasons let me make a remark of pedagogical interest. If eqn (12.4) is true, then

$$K' = \exp\left\{\frac{ie}{\hbar c}[\phi(b) - \phi(a)]\right\} K$$

and we immediately note that a gauge transformation induces only a change of phase in the propagator. The demonstration of gauge-invariance is clumsy to make when working with the Schrödinger equation but as we see here it falls right out of the path-integral.

The first subtlety associated with eqn (12.4) is that because most of the paths γ are very rough the integral is not a Riemann integral, but rather an Itô integral. Specifically, let ϕ be smooth, t_i be a mesh on the interval $[0, T]$, $x(t)$ be the curve γ, $x_i = x(t_i)$, $x_0 = a$, $x_{N+1} = b$, $0 \le \theta \le 1$ and $u_{\theta i} =$

$x_i + \theta(x_{i+1} - x_i)$; then

$$\phi(b) - \phi(a) = \sum_{j=0}^{N} [\phi(x_{j+1}) - \phi(x_j)]$$

$$= \sum_{j=0}^{N} [\phi'(u_{\theta j})(x_{j+1} - x_j) + \tfrac{1}{2}\phi''(u_{\theta j})x_{j+1} - x_j)^2(1 - 2\theta) +$$
$$+ O((x_{j+1} - x_j)^3)]. \quad (12.5)$$

Because for the path-integral, as for the Weiner integral, $(x_{i+1} - x_i)^2 = O(t_{i+1} - t_i)$, the second term in the sum in eqn (12.5) will contribute unless $\theta = \tfrac{1}{2}$. Since we do *not* want this term to contribute, we have the rule: evaluate velocity-dependent terms at the mid-point of the interval. Feynman knew this in 1948 (at least), but there are still papers today that get it wrong. There is a similar phenomenon when the kinetic energy is space- or time-dependent. The consequences of this were examined by DeWitt [5] and by Edwards and Gulyaev [6] but here especially there are hosts of papers where the error in the paper can be traced to neglect of this point.

Returning now to eqn (12.4) we observe that even though $\nabla \phi$ is locally a gradient, because M is not simply connected, the integral

$$\int_\gamma \nabla \phi \cdot d\mathbf{x} \quad (12.6)$$

can take different values for the different paths γ. In fact, we can be certain that (12.6) will take the same value for two paths only if they are continuously deformable into one another, i.e. they fall in the same homotopy equivalence class of paths. In this way matters of homotopy theory enter path-integrals and quantum mechanics in a natural way. We again take $b = a$ for simplicity, so that the equivalence classes mentioned above are just the elements of the fundamental group $\pi(M)$. It follows that the integral (12.6) depends only on the equivalence class of γ. If $\gamma \in g \in \pi$, we define $\nu(g)$ to be

$$\nu(g) = \frac{e}{\hbar c} \int_\gamma \nabla \phi \cdot d\mathbf{x}; \quad (12.7)$$

then the sum in eqn (12.3) can be broken into a separate sum over g and then over $\gamma \in g$,

$$K' = \sum_{g \in \pi} e^{i\nu(g)} \sum_{\gamma \in g} e^{iS(\gamma)/\hbar}. \quad (12.8)$$

It is also obvious that

$$\nu(g_1) + \nu(g_2) = \nu(g_1 g_2), \quad (12.9)$$

since the equivalence class of the composition of two paths is the product of their individual classes (by definition) and

$$\int_{\gamma_1 \circ \gamma_2} \nabla\phi \cdot d\mathbf{x} = \int_{\gamma_1} \nabla\phi \cdot d\mathbf{x} + \int_{\gamma_2} \nabla\phi \cdot d\mathbf{x}.$$

We have thus shown that *if* the propagator is one that arises from a gauge transformation on a propagator with no relative phases, then the new phases have the representation property. To get the stronger result, namely, that (12.8) with ν satisfying (12.9) is the most general form of K, it seems you have to know something about propagators, and in particular about their short-time behaviour.

Once you know how to do a path-integral on a multiply connected space there are a number of interesting problems you can attack. We shall discuss (separately) spin and statistics.

One way to understand spin is to think of the spinning particle as having some extension and rotating in some way. In fact, to do a path-integral for this system you need say very little about the model—only that the coordinate space for the internal motion is the group manifold of $SO(3)$, which is a minimal assumption for any model which can yield conjugate angular-momentum variables. The fundamental group of this space is Z_2 which has two representations. There are therefore two kinds of propagators, which in fact correspond to the propagators of half-integral and integral spin [1].

The author does not feel that we are really justified in using this result as a basis for physical speculations, especially since the relativistic theory [7] is not in very good shape. Nevertheless, it has seemed to me that some of the strictures against thinking of spin as something spinning have been removed. A rotating object *can* give rise to a multivalued wave function when we know how to add amplitudes properly.

The propagator for motion on SO(3) turns out to have a closed form expression which is simply related to the Jacobi theta function. Furthermore, this expression can be obtained by summing over the classical paths (those satisfying the classical equations of motion) only. Dowker [8] has shown that this result is not an isolated one and that geodesic motion on a large class of manifolds exhibits both the foregoing characteristics. Other results dealing with general questions about the classical paths (or the WKB approximation) are to be found in [9]–[16].

Another situation where multivaluedness of the wave functions plays a vital role is that of quantum statistics. Here Laidlaw and DeWitt [4] introduced multiply connected spaces in the following way.

Ordinarily if we have a pair of identical particles we assign them individual coordinates, but then knowing that the Hamiltonian is symmetric in these coordinates we symmetrize (or antisymmetrize) the wave function. It would be more satisfying, however, not merely to ensure that the wave function be symmetric (or antisymmetric) in the coordinate pairs $(\mathbf{x}, \mathbf{y})$ and $(\mathbf{y}, \mathbf{x})$, but to go one step further and say that (**x**, **y**) and (**y**, **x**) are the same point. That is, if we define the equivalent relation

$$(\mathbf{x}, \mathbf{y}) \sim (\mathbf{w}, \mathbf{z}) \quad \text{if} \quad \mathbf{x} = \mathbf{z} \quad \text{and} \quad \mathbf{y} = \mathbf{w},$$

then the proper coordinate space for a pair of *identical* particles is not $\mathbf{R}^3 \times \mathbf{R}^3$ but $\mathbf{R}^3 \times \mathbf{R}^3$ modulo this equivalence relation.

Unfortunately this eminently logical structure needs just a little bit of patching (or scraping) before the space has a fundamental group, which leads to interesting conclusions. What must be done is to take away the points of coincidence $\{(\mathbf{x}, \mathbf{x})\}$. Then, generalizing to N particles, the coordinate space is $\{[\mathbf{R}^{3N} - (\text{coincidences of 2 or more points})]$ modulo the (pairwise) equivalence relation$\}$. The fundamental group of this space is S_N, the permutation group on N objects. S_N has but two one-dimensional representations corresponding to fermions and bosons. Hence, the requirement that the propagator involve only one-dimensional representations forbids the appearance of parastatistics.

I have discussed spin and statistics separately but there is a question which must now be asked. Will the common language developed for handling these two phenomena lead to any new understanding of the uncanny relation between them? The author has nurtured a hope in this direction but is not yet sure the language is equal to the task. Specifically a spin-statistics theorem seems to require a relativistic setting, and as mentioned earlier the relativistic spin theory is not well understood.

12.2. Caustics in electron optics

This subject, like the preceding one, is best discussed in terms of sums over paths rather than the limit of many iterated integrals. In other respects, however, they are very different. We shall examine the limit $\hbar \to 0$ and evaluate only the largest term in the propagator in this asymptotic limit. Near the focal points of the classical motion the propagator becomes large and many of the results we obtain parallel and extend those of classical optics at a caustic.

Consider a spinless non-relativistic particle moving in one or more dimensions with Lagrangian $L = \frac{1}{2}m\mathbf{v}^2 + V(x)$. The propagator is given by (same notation as before)

$$K(b, a; T) = \lim_{N \to \infty} K^{(N)}(b, a; T),$$

$$K^{(N)}(b, a; T) = \left(\frac{m}{2\pi i \hbar \varepsilon}\right)^{(N+1)/2} \int dx_1 \ldots dx_N \exp[iS^{(N)}(x_1, \ldots, x_N)/\hbar],$$

$$S^{(N)}(x_1, \ldots, x_N) = \sum_{j=0}^{N} \varepsilon \left[\frac{m}{2}\left(\frac{x_{j+1} - x_j}{\varepsilon}\right)^2 - V(x_j)\right],$$

$$x_{N+1} = b, \; x_0 = a, \; \varepsilon = T/(N+1).$$

The method of stationary phase is applied to evaluate $K^{(N)}$. This leads to a finite difference equation for the extremum of $S^{(N)}$ which in the limit $N \to \infty$ is just the Euler–Lagrange equation. Let $\bar{x}(t)$ be a solution to this equation (i.e. a classical path) and let $\eta(t) = x(t) - \bar{x}(t)$, so that $\eta(0) = \eta(T) = 0$ for those paths that enter the path-integral. Then

$$K(b, a; T) = e^{iS(\bar{x})/\hbar} \int d\eta(.) \, e^{i[S(\bar{x}+\eta) - S(\bar{x})]/\hbar}. \tag{12.10}$$

We expand $S(\bar{x} + \eta)$ in powers of η,

$$S(\bar{x}+\eta) - S(\bar{x}) = \sum_{k=2}^{\infty} \delta^k S$$

$$= \frac{1}{2} \int_0^T dt \left\{ m\eta^2 - \frac{\partial^2 V}{\partial x^2}[\bar{x}(t)]\eta^2 \right\} - \sum_{k=3}^{\infty} \frac{1}{k!} \int_0^T \frac{\partial^k V}{\partial x^k} \eta^k \, dt.$$

We shall now consider how the standard results, given say by Montroll [17] or Gelfand and Yaglom [18] are obtained. If there are no terms $\delta^k S$, $k \geq 3$, then K can be calculated in closed form giving the answer

$$K(b, a; T) = \sqrt{\left[\frac{m}{2\pi i\hbar f(T)}\right]} e^{iS(\bar{x})/\hbar}, \tag{12.11}$$

where f depends on a and b also and satisfies

$$\frac{\partial^2 f}{\partial t^2}(t) + \frac{\partial^2 V}{\partial x^2}[\bar{x}(t)]f(t) = 0$$

$$f(0) = 0, \quad \frac{\partial f}{\partial t}(0) = 1. \tag{12.12}$$

This formula can be obtained from $\delta^2 S$ for finite N

$$\delta^2 S = \eta^T \cdot \sigma \eta,$$

with $(\eta)_i = \eta(t_i)$, $i = 1, \ldots, N$, and

$$\sigma = \frac{m}{2\varepsilon} \begin{bmatrix} 2 & -1 & & & 0 \\ -1 & 2 & \cdot & & \\ & \cdot & \cdot & \cdot & \\ & & \cdot & \cdot & -1 \\ 0 & & & -1 & 2 \end{bmatrix} - \frac{\varepsilon}{2} \begin{bmatrix} V_1^{(2)} & & & 0 \\ & \cdot & & \\ & & \cdot & \\ 0 & & & V_N^{(2)} \end{bmatrix}, \tag{12.13}$$

where $V_j^{(2)} \equiv \partial^2 V[x(t_j)]/\partial x^2$. The integral in (12.10) is a Gaussian and f is, up to εs, just the limit as $N \to \infty$ of the determinant of σ. Eqn (12.12) is the Jacobi equation and f is a Jacobi field. A classical trajectory out of the point a

will be a minimum of the action until at least one of the Jacobi fields along it vanishes. Points along the trajectory where a Jacobi field vanishes are conjugate points or focal points of a, and after the first of these a trajectory is a saddle-point of S [19, 20].

We shall first discuss why, when terms $\delta^k S$, $k \geq 3$, are present, these are of smaller order in $\hbar$ and the expression (12.11) is still the leading contribution asymptotically as $\hbar \to 0$. Then finally I will get to the point of the story, which is what happens near points where f vanishes and how the higher-order terms $\delta^k S$, $k \geq 3$, must be considered in the neighbourhood of these points.

We change integration variables by diagonalizing $\delta^2 S$. This can be done by keeping all the εs, but it is easier to work formally. The eigenvalue equation for $\delta^2 S$ is

$$m\frac{d^2 \psi_k}{dt^2} + \frac{\partial^2 V[(\bar{x}(t)]}{\partial x^2}\psi_k + \lambda_k \psi_k = 0,$$

$$\psi_k(0) = \psi_k(T) = 0, \qquad \int_0^T \psi_k(t)\psi_j(t)\,dt = \delta_{kj}. \tag{12.14}$$

This Sturm–Liouville equation provides a complete set of functions on the interval $[0, T]$ and $\eta(t)$ can be expanded

$$\eta(t) = \sum_{k=1}^{\infty} a_k \psi_k(t). \tag{12.15}$$

(For finite N, the column vector η is expanded in the eigenvectors of σ.) In the coordinates $\{a_k\}$ the integral over $\eta(t)$ becomes

$$K(b, a; T) = e^{iS(\bar{x})/\hbar} \int \left(\prod_1^{\infty} da_k\right) \exp\left\{\frac{i}{\hbar}\left[\tfrac{1}{2}\sum \lambda_k a_k^2 + \sum_{j=3}^{\infty} \delta^j S\right]\right\}. \tag{12.16}$$

It is instructive to carry out the above steps keeping all the εs—it is a chance to go through a nice derivation of Sturm–Liouville theory. For example, the fact that $\lambda_k \sim k^2$ as $k \to \infty$ can be obtained from the known exact eigenvalues of the Jacobi matrix

$$\begin{bmatrix} 2 & -1 & & & 0 \\ -1 & 2 & \cdot & & \\ & \cdot & \cdot & \cdot & \\ & & \cdot & \cdot & -1 \\ 0 & & & -1 & 2 \end{bmatrix}$$

Let us count powers of $\hbar$ to justify dropping $\delta^k S$, $k \geq 3$. The integrand in (12.16) is stationary for $a_j = 0$, $j = 1, 2, \ldots$. Assume for now that this is the only stationary point. Regions of the a_j integration where $(a_j^2 \lambda_j/\hbar)$ is large will give negligible contributions to the integral; i.e. to lowest order in $\hbar$ the

two results of adding path amplitudes

region of integration in each a_j is effectively

$$\frac{1}{\hbar}\lambda_j a_j^2 \leq 1. \tag{12.17}$$

We are going to use the inequality (12.17) again and again, but it is not really true that truncation of all integrals in accordance with (12.17) would leave lowest order results unchanged. It would be more accurate to require $a_j^2 \lambda_j \leq \hbar |\log \hbar|$ but, since we will always be counting *powers* of $\hbar$, logarithmic terms will not matter. Note by the way that (12.17) implies that $a_j \sim 1/j$ as $j \to \infty$, which is just the right behaviour for coefficients of a function η without derivatives.

The real problem in using (12.17) arises from the infinity of integration variables. To study this one can go back to finite N and examine the collective error for all N arising from the stationary-phase approximation. The error for each a_k, $k = 1, \ldots, N$, will go to zero with $\hbar$, but this error must also converge as $N \to \infty$. If we were dealing with the Wiener integral instead of the path-integral, we would be in good shape. Schilder [21] has proved the validity of the Laplace method for integrals of the kind we are considering (but $i \to -1$, of course) and he and Varadhan [22]–[24] have applied these results to a number of differential equations. But for the path-integral, i.e. for the method of stationary phase with infinitely many variables, there are, as far as the author knows, no rigorous results. Nevertheless, the author has no doubt that a näive application of the method of stationary phase is correct since this is just the WKB approximation.

Formula (12.11) breaks down at and near focal points, for then a, b, and T are such that there is a solution of eqn (12.12) with $f(T) = 0$. But by eqn (12.15) this means that (at least) one of the eigenvalues of $\delta^2 S$ is zero. Assume that only one eigenvalue goes to zero and that this is λ_1. Referring to eqn (12.17) we note that a_1 will now take large values and in fact its size will be fixed by the higher powers of a_1 that appear in $\sum_{k \geq 3} \delta^k S$. If the coefficient of a_1^3 does not vanish, the analogue of eqn (12.17) is

$$\frac{ua_1^3}{\hbar} \lesssim 1, \quad u = -\frac{1}{2}\int_0^T \frac{\partial^3 V}{\partial x^3}[\bar{x}(t)][\psi_1(t)]^3 \, dt. \tag{12.18}$$

The order of a_1 is $\hbar^{\frac{1}{3}}$ rather than $\hbar^{\frac{1}{2}}$. It is also evident that λ_1 need not be strictly zero for the cubic term to be significant, and that the critical region is $\lambda_1 \lesssim \hbar^{\frac{1}{3}}$. The integrals over a_j, $j \neq 1$, are not affected and, counting powers of $\hbar$ (e.g. $\hbar^{-1} a_1^2 a_k \sim \hbar^{\frac{1}{6}} \to 0$), these can be dropped. The approximation to the propagator is therefore

$$K(b, a; T) = e^{iS(\bar{x})/\hbar}\sqrt{\left(\frac{-m\lambda_1}{2\pi^2\hbar^2 2f}\right)}\int_{-\infty}^{\infty} da_1 \exp\left\{\frac{i}{\hbar}(\tfrac{1}{2}\lambda_1 a_1^2 + \tfrac{1}{3}ua_1^3)\right\}, \tag{12.19}$$

where the quotient $(\lambda_1/2f)$ is just a way of writing $\prod_{j=2}^{\infty}(\lambda_j)$ ($\times$ appropriate factors of ε, 2, etc.).

The propagator of eqn (12.19), with its strong resemblance to an Airy function, is close enough to describing a caustic to deserve some comment; however, with a slight change in formulation it is possible to get exactly the Airy function. Suppose a point b is conjugate to a, where $\bar{x}(0) = 0$, $\bar{x}(T) = b$, and $\bar{x}(t)$ is a classical path. Then there exists a function, say, $\psi_1(t)$, which satisfies eqn (12.12) with $\lambda_1 = 0$ and with $\partial^2 V/\partial x^2$ evaluated along this $\bar{x}(t)$. If we are interested in the propagator

$$K(b+\Delta b, a+\Delta a; T)$$

for points near a and b, we do *not* expand about the classical path from $(a+\Delta a)$ to $(b+\Delta b)$, but rather about $\bar{x}$ from a to b. This expansion is preferred because there may not be any classical path from $(a+\Delta a)$ to $(b+\Delta b)$ in time T. Indeed the disappearance of a path is a characteristic phenomenon at a caustic (we shall see this in more detail shortly). Let ψ_k and λ_k be the eigenvectors and eigenvalues defined with respect to $\bar{x}$. Then it is easy to get the formal result

$$K(b+\Delta b, a; T) \sim \exp\{i[S(\bar{x}) - m\dot{\bar{x}}(T)\Delta b]/\hbar\} \int da_1 \exp\left\{\frac{i}{\hbar}[-\mathbf{w}\cdot\Delta\mathbf{b} + ua_1^3/3]\right\}, \quad (12.20)$$

where $\mathbf{w}$ is a vector proportional to $\dot{\psi}_1(T)$ (and for simplicity Δa is taken to be zero). The propagator now does contain an Airy function. To recall some properties of this function, let

$$I(\nu, c) = \int_{-\infty}^{\infty} \exp[i\nu(\tfrac{1}{3}x^3 - cx)]\,dx, \quad (12.21)$$

so that in terms of the Airy function $\mathrm{Ai}(t)$,

$$I(\nu, c) = \frac{2\pi}{\nu^{\frac{1}{3}}} \mathrm{Ai}(-\nu^{\frac{2}{3}}c). \quad (12.22)$$

For $\nu \to \infty$ the leading contribution to I is found by making the argument of the exponential in (12.21) stationary, i.e. by solving the equation

$$x^2 - c = 0. \quad (12.23)$$

The sign of c thus has a dramatic effect on I and asymptotically

$$I(\nu, c) \sim \begin{cases} \sqrt{\left(\dfrac{\pi}{i\nu\sqrt{(c)}}\right)} \exp\left(\dfrac{2i}{3}\nu c^{\frac{3}{2}}\right) + \sqrt{\left(\dfrac{-\pi}{i\nu\sqrt{(c)}}\right)} \exp\left(-\dfrac{2i}{3}\nu c^{\frac{3}{2}}\right) & \text{for } c>0 \\[2ex] \sqrt{\left(\dfrac{\pi}{\nu\sqrt{(-c)}}\right)} \exp[-\tfrac{2}{3}\nu(-c)^{\frac{3}{2}}] & \text{for } c<0. \end{cases} \quad (12.24)$$

The propagator in eqn (12.20) therefore depends on the sign of

$$c = \frac{\mathbf{w} \cdot \Delta \mathbf{b}}{u}.$$

For $c < 0$ we are in the 'shadow' region of exponential ($\sim |c|^{\frac{3}{2}}$) decay, while the region for which $c > 0$ is illuminated. This behaviour is well known from electromagnetic theory ([25],[26]). It is also evident that $\psi_1(T)$ is the normal to the caustic.

Eqn (12.24) is a good example of an expansion that is *not* uniform in v and c. If c passes through zero from above, the critical points of eqn (12.23) will coalesce and move onto the imaginary axis. The importance of this issue was emphasized by Erdelyi [27] and for a uniform expansion we must go back to the Airy function as in eqn (12.22). The region in which the uniform expansion must be used is of size $vc^{\frac{3}{2}} \sim 1$ which in our case means $\Delta b \sim \hbar^{\frac{3}{2}}$. The existence or non-existence of a root of eqn (12.23) is of direct physical significance. This is exactly the question of whether or not there exists a path (near $\bar{x}$) from a to $b + \Delta b$ in time T. In the earlier discussion (before eqn (12.17)) we may have had misgivings about the assumption that there were no stationary points for a_j different from zero. Now we can see how that is handled. If there is a stationary point for very large a_j then this simply represents another classical path distant from the one we are expanding about and its contribution is added separately. On the other hand, if there is a stationary value of $\{a_j\}$ near zero it can cause trouble—but this is just the trouble we have been examining systematically, since the coalescence of paths is accompanied by a vanishing of some eigenvalue of $\delta^2 S$.

We shall now consider the vanishing of

$$u = -\frac{1}{2} \int_0^T \frac{\partial^3 V}{\partial x^3}[\bar{x}(t)][\psi_1(t)]^3 \, dt$$

and also the vanishing of more than one eigenvalue λ_1 at the same point. In the first case the size of a_1 will be determined by

$$\hbar^{-1} a_1^4 \sim 1 \quad \text{or} \quad a_1 \sim \hbar^{\frac{1}{4}}$$

(if the coefficient of a_1^4 does not vanish) and the critical region is $\lambda_1 \leqslant \hbar^{\frac{1}{2}}$. The propagator can be brought to the form

$$K(b + \Delta b, a; T) \sim e^{iS(\bar{x})/\hbar} \int_{-\infty}^{\infty} da_1 \exp\left[\frac{i}{\hbar}\left(\frac{a_1^4}{4} + \frac{xa_1^2}{2} + ya_1\right)\right]. \quad (12.25)$$

The quantities x and y depend on Δb and vanish when $\Delta b = 0$. The point b is even more brightly illuminated than in previous cases. Moreover, in some directions out of the point b the polynomial

$$P(a_1) = \frac{a_1^4}{4} + \frac{xa_1^2}{2} + ya_1 \quad (12.26)$$

has points where it is stationary and also its second derivative vanishes. This means that by a shift in variable we are again on a caustic but one of lower order, one in which a cubic polynomial determines the behaviour. The equation of this caustic is obtained by finding those (x, y) for which the equations

$$\frac{\mathrm{d}P}{\mathrm{d}a_1} = a_1^3 + xa_1 + y = 0,$$

$$\frac{\mathrm{d}^2 P}{\mathrm{d}a_1^2} = 3a_1^2 + x = 0 \qquad (12.27)$$

can be solved simultaneously. The required relation between x and y is

$$\frac{x^3}{27} + \frac{y^2}{4} = 0,$$

which is a cusp. x and y are definite calculated functions of Δb, y being a coordinate in the direction $\dot{\psi}_1(T)$, and x a coordinate in a perpendicular direction. Thus, this particular caustic assumes a specific form, that of the cusp.

Suppose at some b, both λ_1 and λ_2 are zero. K will then involve a double integral and it is obvious that terms in a_1 and a_2 will appear with coefficients related to $\Delta \mathbf{b} \cdot \dot{\psi}_1(T)$ and $\Delta \mathbf{b} \cdot \dot{\psi}_2(T)$. But this is not all. The component of $\Delta \mathbf{b}$ along the direction $\dot{\psi}_1(T) \times \dot{\psi}_2(T)$ will also control the development of the caustic due to terms from $\delta^3 S$. The propagator can therefore be written

$$K \sim e^{iS/\hbar} \int \mathrm{d}a_1 \, \mathrm{d}a_2 \, e^{iP/\hbar}, \qquad (12.28)$$

where P can be brought to the form

$$P(a_1, a_2) = a_1^3 + a_2^3 - xa_1 - ya_2 - za_1 a_2. \qquad (12.29)$$

We can in this way obtain an entire class of functions giving a uniform description of the wave function near a caustic. Those of the kind presented in eqn (12.25) have been given the name 'generalized Airy functions' [25], [26] and it would seem the propagator of eqn (12.28) with a polynomial in several variables would warrant such a designation also.

The picture that emerges is that a certain polynominal (e.g. (12.20) or (12.26)) by virtue of its appearing in an integral with a large parameter governs the form taken by the caustic. Specifically, several of the coefficients of the lower powers in the polynomal are functions of position. The location of the caustic corresponds to the values of the coefficients for which the polynomial considered as a potential would be structurally unstable. Thus, the equation of the cusp was found from the polynomial (12.26) via eqns (12.27).

Our results here conform to the general scheme predicted by Thom [28] and the caustics of electron optics have been found to be what he would call catastrophes. In fact, in his book he has a number of photographs of electromagnetic caustics to illustrate several of the possible forms of a catastrophe.

As is usual in applying Thom's ideas the description in terms of catastrophes is just the first step. Once you allow that the over-all form of the system is accounted for by the simple polynomials, if you want details of the behaviour, you must put in the details of the system. The variation of K very near the caustic—by analogy with statistical mechanics this might be called critical behaviour—is one such detail that requires us to put the polynomial into a specific context, and not just know that there is a polynomial.

Nevertheless, despite the author's reservations about the ability of catastrophe theory to answer the detailed questions that scientists are prone to ask, it must be considered something of a triumph that general predictions about form can be made with so little input.

In conclusion some remarks of a pessimistic nature must be made on why the Laplace method fails for some of the path-integrals encountered in statistical mechanics. A functional integral of great interest is

$$Z = \int dM(.) \exp[-\beta F(M)],$$

$$F(M) = \int_V d^n r\{[M(r)]^4 + a(T)[M(r)]^2 - B(r)M(r_1) + c(\nabla M)^2\},$$

where $M(r)$ is the order parameter, say magnetization, for a system in a volume V in n-dimensional space. For temperature T_c such that $a(T_c) = 0$, $\delta^2 F$ has a zero eigenvalue and the temptation is to use the Laplace method with the volume V being the parameter that becomes large. Unfortunately, trouble is encountered because this very largeness leads to the existence of other eigenvalues of $\delta^2 F$ which crowd close to zero as $V \to \infty$. Specifically, if F is fourier analysed the coefficient of M_k^2 is $a(T) + k^2$. From our previous results the critical region for $a(T)$ is $a(T) \leq V^{-\frac{1}{2}}$. But as $V \to \infty$ the discrete ks (due to the boundary conditions) become as small as $V^{-1/n}$. In order for there to be a critical region in which one mode M_1 (analogous to ψ_1) dominates, it is necessary for $a(T) \ll k_{\min}^2$ which will be the case only for $n > 4$. This does not prove that the Laplace method does work for $n > 4$, since there are still many modes clustered near zero even though they may not be near enough to destroy our first estimates. Nevertheless, the dimension 4 is picked out in such a way that beyond 4 things seem to get simpler [29].

Acknowledgements

Some of the work reported here is based on the dissertations of D. MacLaughlin [30] and S. Coyne [31]. Coyne in particular examined the asymptotic expansion for the propagator in the limit $\hbar \to 0$. I have also benefited from discussions with M. Donsker, E. Lieb, and R. Shtokhamer.

This work has been supported by U.S. Army Research Office, Durham, North Carolina.

References

1. SCHULMAN, L. S. *Phys. Rev.* **176**, 1558 (1968).
2. SCHULMAN, L. S. *Phys. Rev.* **188**, 1139 (1969).
3. SCHULMAN, L. S. *J. Math. Phys.* **12**, 304 (1971).
4. LAIDLAW, M. G. G. and DE WITT, C. M. *Phys. Rev.* **D3**, 1375 (1971).
5. DE WITT, B. *Rev. mod. Phys.* **29**, 377 (1957).
6. EDWARDS, S. F. and GULYAEV, Y. V. *Proc. R. Soc.* **A279**, 299 (1964).
7. SCHULMAN, L. S. *Nucl. Phys.* **B18**, 595 (1970).
8. DOWKER, J. S. *J. Phys. A* **3**, 451 (1970) and *Ann. Phys. N.Y.* **62**, 361 (1971).
9. NORCLIFFE, A. and PERCIVAL, I. C. *J. Phys. B* **1**, 774 (1968).
10. NORCLIFFE, A. and PERCIVAL, I. C. *J. Phys. B* **1**, 784 (1968).
11. NORCLIFFE, A. and PERCIVAL, I. C., and ROBERTS, M. J. *J. Phys. B* **2**, 578 (1969).
12. NORCLIFFE, A., PERCIVAL, I. C., and ROBERTS, M. J. *J. Phys. B* **2**, 590 (1969).
13. NORCLIFFE, A. *J. Phys. B* **4**, 143 (1971).
14. BALIAN, R. and BLOCH, C. Solution of the Schrodinger equation in terms of classical paths. *Ann. Phys. N.Y.* (To appear.)
15. BERRY, M. V. and MOUNT, K. E. *Rep. Prog. Phys.* **35**, 315 (1972).
16. BABICH, V. M. and DANILOV, YU. P. In *Seminars in mathematics*, Vol. 15; *Mathematical problems in wave propagation theory*, Part II (ed. V. M. Babich), p. 23. Consultants Bureau, New York (1971).
17. MONTROLL, E. W. *Commun. pure appl. Math.* **5**, 415 (1952).
18. GELFAND, I. M. and YAGLOM, A. M. *J. Math. Phys.* **1**, 48 (1960).
19. MILNOR, J. *Morse theory*. Ann. Math. Ser. No. 51, Princeton University Press (1963).
20. POSTNIKOV, M. M. *The variational theory of geodesics*. Saunders, Philadelphia (1967).
21. SCHILDER, M. *Trans. Am. math. Soc.* **125**, 63 (1966).
22. VARADHAN, S. R. S. *Commun. pure appl. Math.* **19**, 261 (1966).
23. VARADHAN, S. R. S. *Commun. pure appl. Math.* **20**, 431 (1967).
24. VARADHAN, S. R. S. *Commun. pure appl. Math.* **20**, 659 (1967).
25. LUDWIG, D. *Commun. appl. Math.* **19**, 215 (1966).
26. BLEISTEIN, N. *J. Math. Mech.* **17**, 533 (1967).
27. ERDELYI, A. In *Symposium on analytic methods in mathematical physics, Indiana University* (1968) (eds R. P. Gilbert and R. G. Newton). Gordon and Breach, New York (1970).
28. THOM, R. *Stabilite structurelle et morphogeneses*. Benjamin, Reading, Mass. (1972).
29. WILSON, K. G. *Phys. Rev. B* **4**, 3184 (1971).
30. MCLAUGHLIN, D. W. The path integral: its approximation in flat space and its representation in curved space. Thesis, Indiana University (1970).
31. COYNE, S. Semi-classical asymptotic evaluation of Feynman path integrals. Thesis, Indiana University (1972).

13. Functional integration and interacting quantum fields

I. E. SEGAL

13.1. Introduction

THE questions of the multiplication of interacting quantum fields lie at the scientific foundations of quantum field theory. What do non-linear local expressions in quantum fields really mean to physicists and mathematicians? How compatible are the physically natural desiderata and the inevitable mathematical limitations on what exists and is analytically viable? For almost 50 years the outstanding character of these questions has blocked the resolution of the existence and properties of quantum fields satisfying a prescribed non-linear equation, such as $\Box \phi = m^2 \phi + \phi^3$ to take a currently popular model equation.

The ϕ^3 term which occurs here has only intuitive meaning. A well-defined mathematical problem arises only when it is given definite mathematical meaning; a straightforward definition fails, and it is clear that a mathematically unconventional approach is called for. But from the standpoint of the over-all scientific importance of quantum field theory, much more than purely mathematical naturalness and cogency are required. We want to know not merely whether a mathematical theory exists which may be identified with the formalism of physical quantum field theory, but whether it has physical cogency. The fecundity of the human mind in matters of mathematical interpretation strongly suggests that we can have complete confidence in a purely mathematical non-linear quantum field theory—short of extraordinary good fortune in the matter of the naturalness and definitiveness of the formulation of the theory—only if it is accompanied by effective methods of computation, a rounded mathematical treatment of the main secondary foundational problems, and quantitative applications subject to experimental confirmation.

We appear still to be a long way from this, though the gaps may close rapidly once the right underlying mathematical theory is available. What will be attempted here is an assessment as of the present moment of this situation, from both the formal standpoint of the 'founding fathers' of quantum field theory—Dirac, Heisenberg, and Pauli—and from the modern mathematical position regarding local partial differential equations. This is indeed a matter of functional integration; in the case of the free field, the Wick-ordered powers are closely related to Wiener's notion of homogeneous chaos, e.g. the general treatment of local partial differential equations, corresponding to interacting fields, seems to involve local functions of distributions which are equivalent to functional integrals.

158   *Functional integration*

At a formal level the formulation of quantum field theory reached an early stationary state in the work of Heisenberg and Pauli, around the end of the 1920s. The original Heisenberg prescription, together with the Dirac extension from a finite set of geometrical variables to an infinite set of field components, was very simply and naturally incorporated in the following way.

To quantize an equation such as
$$\Box \phi = m^2 \phi + p'(\phi),$$
where p is a given polynomial, we express the classical Hamiltonian as
$$H_{cl} = \int (\nabla \phi)^2 \, d\mathbf{x} + m^2 \int \phi^2 \, d\mathbf{x} + \int \dot{\phi}^2 \, d\mathbf{x} + \int p(\phi) \, d\mathbf{x};$$
we then replace the classical field ϕ and its first time derivative $\dot{\phi}$ by the corresponding operator-valued quantized fields $\hat{\phi}$ and $\dot{\hat{\phi}}$ satisfying the equal-time commutation relations
$$[\hat{\phi}(\mathbf{x}, t), \hat{\phi}(\mathbf{y}, t)] = 0 = [\dot{\hat{\phi}}(\mathbf{x}, t), \dot{\hat{\phi}}(\mathbf{y}, t)];$$
$$[\hat{\phi}(\mathbf{x}, t), \dot{\hat{\phi}}(\mathbf{y}, t)] = i\delta(\mathbf{x} - \mathbf{y})$$
in the original differential equation; or equivalently
$$\hat{\phi}(\mathbf{x}, t) = e^{-itH} \hat{\phi}(\mathbf{x}, 0) e^{itH},$$
where the quantum-mechanical Hamiltonian H is obtained from the classical one by the substitutions $\phi \to \hat{\phi}$, $\dot{\phi} \to \dot{\hat{\phi}}$. This process is independent of the choice of t, as is H_{cl}, and is effectively independent of the Lorentz frame, although of course the Hamiltonians are frame-dependent.

In other terms, the Hamiltonian of the interacting quantum field ϕ_{int}, H_{qm}, is expressible as a formally well-defined polynomial in the interacting quantum field.
$$H_{qm}(\phi_{int}) = H_{free}(\phi_{int}) + H_{int}(\phi_{int}),$$
consisting of 'free' and 'interacting' components, where 'free' and 'interacting' refer to the formal expression, and not to the field substituted in the expressions, this field being the interacting quantum field in both cases. In the free case, $p \equiv 0$, this equation gives the free Hamiltonian $H_{qm}(\phi_{free})$ as $H_{free}(\phi_{free})$. When an in-field ϕ_{in} exists, it results that
$$H_{free}(\phi_{in}) = H_{free}(\phi_{int}) + H_{int}(\phi_{int}),$$
both sides of this equation being time-dependent.

Before about 1940, it was devoutly hoped and believed that the canonical commutation relations (CCRs) had a unique irreducible representation, so that at any time the Cauchy data for the free and interacting fields could be identified as unitarily equivalent. This folklore did not survive the decade following the Second World War, which showed that the interacting field was sui generis, and could be identified with the free field only asymptotically, at time $\pm\infty$, in the absence of bound states. As a consequence, the Heisenberg–Pauli Hamiltonian $H_{qm}(\phi_{int})$ was inherently a rather involved object, underneath its deceptive formal simplicity. For in order to propagate the interacting field temporally, the interacting field at one time at least had to be known; but to know that the interacting field had been obtained, either a solution of the partial differential equation of motion or the $H_{qm}(\phi_{int})$ was needed. The equation for interacting field was thus totally implicit, and in particular the relevant representation of the CCRs (at a sharp time) could not be given *a priori*, but had to be solved for, along with the rest of the structure of the interacting field.

The Nelson *Ansatz* [3] of simplifying the Hamiltonian by substituting the free field in place of the interacting field in the expression for the total Hamiltonian, $H = H_{free}(\phi_{int}) + H_{int}(\phi_{int})$, was mathematically unexceptionable as a means of developing a repertoire of exploratory models, although physically the substitution is *a priori* no less inappropriate than it would be classically. In either case the resulting Hamiltonian is time-dependent, and is at best merely asymptotic to the true H as $t \to \pm\infty$. This physical point was apparently unappreciated in some of the enthusiastic early development of concrete models, in particular the work of Glimm and Jaffe, as well as in the lucid exposition of this work by Hepp [1], in which the *Ansatz* is presented as fundamental physics. It is extraordinary that we are nevertheless led ultimately to a field satisfying a local relativistic partial differential equation, although not the original one. Counter terms rear up, just as they do in perturbation theory, and obscure the situation: we still do not know if we can solve any *given* non-linear quantized equation although the existence of equations admitting solutions is assured. On the other hand, when the theory is redeveloped in terms of the interaction representation (cf. [5]), it is closely related to the formulation of the S-matrix as the time-ordered exponential of the *original* interaction Lagrangian, and so would give a direct answer to a physically interesting question (modulo the interest of the fundamental differential equation) if the cogent existence of the infinite space–time limit could be shown.

It is the Heisenberg dynamics which has been considered physically more fundamental than the interaction–representation dynamics, which is much more simply analysed. The unrenormalized C^*-dynamics or the unitary dynamics with a spatial cut-off, for two-dimensional scalar fields with non-negative polynomial self-interaction was well established 5 years ago, as were the unitary dynamics with a spatial cut-off. The main development of the past 5 years has been the infinite-volume limit. Unfortunately, this has done little to clarify the fundamental questions of the existence, unicity, and regularity of the quantized solutions to given non-linear relativistic equation, and their temporal asymptotics. The physically still more crucial questions of the nature of the analogous four-dimensional fields and of treatments that do not make essential use of the scalar character of the field (as many of the developments of the past 5 years appear to) likewise cry out for serious mathematical attention. These are probably very difficult, although the author believes currently accessible questions. Before we can begin, however, the precise meaning and nature of the multiplication of interacting quantum fields must evidently be elucidated. It is this preliminary task which will next be considered.

13.2. Renormalized powers of quantized distributions

Let S denote space. Let a 'clothed Weyl system over S' or 'canonical quantum process over S' be defined as a triple $(\mathbf{K}, W, v)$, where $\mathbf{K}$ is a complex Hilbert space; W is a Weyl system over a class of pairs of real measurable functions on S, *relative to given measure m*, with antisymmetric form

$$A[(f_1, g_1), (f_2, f_2)] = \int (f_1 g_2 - f_2 g_1) \, dm,$$

and v is a given unit vector, for short the 'vacuum', in $\mathbf{K}$.

This definition is both too formal, and not formal enough; there is no perfect way to lay it out for all possible applications. What we are thinking of is something like a quantized scalar field, at a sharp time, inclusive of the vector state-space $\mathbf{K}$, the field and its first time derivative in the exponentiated form given by W, and the field vacuum v. It is, of course, no essential loss of generality to take the vacuum as a normalizable vector, since the Hilbert space can always be altered to make it so; still, there are cases where that may not be convenient. For simplicity and reasonable comprehensiveness regarding the basic concepts, however, we shall use the formal arrangement indicated.

To motivate our investigations, let R denote the infinitesimal form of W, i.e. $R(z)$ is the self-adjoint operator given by Stone's theorem such that $W(tz) = \exp[itR(z)]$ for all real t, where z varies over the given class C. If z is the pair (f, g), the $R(z)$ is the closure of the sum of $R[(f, 0)]$ with $R[(0, g)]$, and we write symbolically

$$R[(f, 0)] \sim \int \phi(x) f(x) \, dx, \qquad R[(0, g)] \sim \int \psi(x) g(x) \, dx,$$
$$[dx \equiv dm(x)].$$

Then $\phi(x)$ and $\psi(x)$ are symbolically 'canonically conjugate operator-valued distributions on S',

$$[\phi(x), \phi(x')] = 0 = [\psi(x), \psi(x')]; \qquad [\phi(x), \psi(x')] = i\delta(x - x')$$

for arbitrary x and x' in S. We may think of $\phi(x)$ as the scalar field, and $\psi(x)$ as its first-time derivative, for definiteness. But in principle the situation is a static one; we have an operator-valued distribution over space, the time being held fixed, and we want to treat what a power of it signifies, or should signify (or, more generally, products of such operator-valued distributions).

We can forget, of course, about the straightforward definition as a limit of powers of functions converging to the distribution in question; this limit does not exist at all, and there is no way simply to fall back on the known definition for strict functions. This may initially be quite upsetting to those brought up in a highly academic mathematical environment; by what authority, they may ask, do we tinker with the concept of power by adding formally infinite terms, and what basis have we for our eventual revised definition? We can only counsel patience and suggest that a reasonable and substantially unique local notion of renormalized power does turn out to exist, and is really not much worse and rather more novel than the concept of a weak solution of a differential equation.

Physicists have learned the hard way that the safest course (not always available, unfortunately) is to rely only on high-school algebra and elementary calculus; and we start out in this spirit. Take the square $\phi(x)^2$ of the field $\phi(x)$, and average with a weight (or test) function f; what intrinsic properties may perhaps be used to characterize mathematically the formal object $\int \phi(x)^2 f(x) \, dx$? If this formal operator is denoted H, then evidently $[H, \phi(y)] = 0$ for all y, but this is very far from characterizing H as a square. These properties, in the usual case, or with the reasonable further assumption that the totality of the $\phi(x)$ generate a maximal commuting set, merely assert that H is in some sense a function of all the $\phi(x)$. However, if we form $[H, \psi(y)]$, a more specific result is obtained: $2i \int \phi(x) \delta(x-y) \, dx = 2i\phi(y)$. Thus, although H is at this state a purely symbolic operator, its commutator with $\psi(y)$ is a perfectly finite one (distribution—theoretically).

Accordingly, we are led to two unexceptable mathematical constraints on the putative operator H, if it exists at all:

$$[H, \phi(y)] = 0, \quad [H, \psi(y)] = 2i\phi(y) \quad (y \in S).$$

It is a definitive mathematical question (apart, perhaps, from regularity hypotheses which we may hope to avoid) whether any such operator H exists. If it does, then why not define it to be the sought for $\int \phi(x)^2 f(x) \, dx$? Of course, these commutation relations leave open a possible additive constant in H; but making the generally satisfied further assumption of irreducibility for the $\phi(x)$ and $\psi(y)$, this is the total extent of the interdeterminacy in H. And even this one additive constant can be determined uniquely if the natural desideratum that the vacuum expectation value of $\phi(x)^2$ should vanish is imposed. (At least in field-theoretic applications it is natural to suppose this, for the expectation value would be a Lorentz-invariant quantity which for large times should approximate the corresponding quantity for the free field, which, however, decays to zero as $|t|$ becomes infinite. Any non-vanishing expectation values would therefore badly clutter up the possibilities for asymptotic convergence of the interacting to a free incoming or outgoing field.)

At first glance it may seem like wishful thinking to hope that such a simple scheme may be effective. The idea is clearly extendable to high powers, etc.; but do non-trivial example exist of operators such as the sought-for H, and are they otherwise entirely reasonable from a quantum field-theoretic standpoint? It is pleasant to be able to report that they do. In the special case of a free field, in particular, we get another definition of the Wick-ordered powers—not that this has any sacred foundational standing, for Wick most modestly and properly put forth his formalism primarily as a means of simple standardization and facilitation of computation, but it should serve to reassure practising physicists that the objects that arise are familiar, reasonable ones, and not exotic abstruse mathematical objects which one would otherwise not anticipate had physical relevance.

The definition by commutators works extremely well for the space-smeared Wick powers, most notably perhaps in proving, and in fact making manifest, the locality of these operators. That is to say, the Wick power $\int :\phi(x)^n: f(x) \, dx$ is clearly seen to be a function of the values of the $\phi(x)$ in the region, say F, in which f is non-vanishing, and thus to commute with the $\psi(y)$ for all values of y outside F. It has also very convenient convergence and formal properties. Briefly, in particular, we can give a simple explicit closed form for the commutation relations between the exponentiated powers $\exp[i \int :\phi(x)^n: f(x) \, dx]$ and the original Weyl operators $W(z)$. These relations uniquely characterize the Wick powers, and since they concern unitary operators, involve no messy or academic domain requirements. This is strictly true only in two space–time dimensions, but relations effectively as good also apply in n space–time dimensions.

The Wick powers are not altogether too simple to be useful for constructive quantum field theory, particularly if we use the interaction representation. The central problem of the theory then becomes that of the *time*-ordered exponentiation of the Wick powers. The author believes that this question has had far less attention that it properly deserves, but has previously made a similar point and will not dwell on it here. If we want to treat the Heisenberg fields—and it may be relatively imprudent, or at least remote from quantitative application, to push the matter at this time—then we are forced into consideration of the meaning of non-linear local functions of the field relative to states other than the free vacuum.

With the experience gained from the analysis of the situation for free fields, no ambiguity appears in the appropriate formulation of renormalized powers, at least in the case in which we wish to require (as certainly one would in two space–time dimensions) that the space-smeared powers should form actual self-adjoint operators and not merely generalized operators.

and interacting quantum fields 163

Specifically, for example, if $H = \int \phi(x)^2 f(x)\,dx$, then $\exp(isH)$ commutes with $\exp[i\int \phi(x)h(x)\,dx]$ for all admissible values of h, and all real values of s, and has as its commutator with $\exp[ir\int \psi(y)g(y)\,dy]$ the value inferred from the bracket it has with $\int \psi(y)g(y)\,dy$. A simple computation shows this to amount to

$$\exp[i\int \psi(y)g(y)\,dy]\exp(isH)\exp[-i\int \psi(y)g(y)\,dy]\exp(-isH)$$
$$= \exp[2is\int \psi(x)f(x)g(x)\,dx]\exp[is\int f(x)g(x)^2\,dx].$$

This relation can determine H only within an additive constant, which is specified by requiring that v lie in the domain of H and that $\langle Hv, v\rangle = 0$, i.e. H has vanishing expectation value.

Once we have treated the square of the field in this way, the road is open to do the same for cubes, basing the extension on the already accomplished theory for squares. More generally, having treated up to and including the $(r-1)$th power in the same vein, the putative operator formally expressible as $H = \int \phi(x)^r f(x)\,dx$ is to be determined by the vanishing of its expectation value, its commutation with $\exp[i\int \phi(x)h(x)\,dx]$ for all values of h, and the relation

$$\exp[i\int \psi(y)g(y)\,dy]\;\exp(isH)\;\exp[-i\int \psi(y)g(y)\,dy]\;\exp(-isH)$$
$$= \exp[i\binom{r}{1})\int \phi(x)^{r-1}f(x)g(x)\,dx]\;\exp[i\binom{r}{2})\int \phi(x)^{r-2}f(x)g(x)^2\,dx$$
$$\ldots \exp[i\binom{r}{r})\int f(x)g(x)^r\,dx].$$

If H exists at all, in the indicated sense, its unicity follows from the postulated irreducibility of the original Weyl system. Thus it is a very well-posed mathematical problem, to what extent the renormalized powers $H = \int \phi(x)^r g(x)\,dx$ exist for any given triple $(\mathbf{K}, W, v)$; naturally the class of weight functions g must be given in advance, but we would not expect this to be of more than technical significance— nor is it. Non-existence of the renormalized powers with a fairly small, conservative class of gs, say those which are smooth and of compact support, implies *a fortiori* such non-existence for any larger class. Conversely, once existence is established for a minimal class of gs, the formulation or determination of maximal algebras of functions g such that the defining equation holds is clear-cut, if perhaps technically involved.

Among the simplest situations in which the existence of the renormalized powers is relevant is that of a field $(\mathbf{K}, W, v)$ in which $(\mathbf{K}, W)$ is that of the free field, but v is not the free vacuum vector (representing physically, on occasion, the ground state of a perturbed field). The situation in this regard extends that described in [6] as follows. *In two space–time dimensions the renormalized powers exist as self-adjoint-operator-valued distributions,* employing, as test functions, say, infinitely differentiable ones of compact support. *In more than two space–time dimensions, the renormalized powers exist as generalized operators, i.e. as sesquilinear forms on a smooth domain,* e.g. that of infinitely differentiable vectors relative to the free Hamiltonian. *In this case the renormalized power exists as an actual point function,* i.e. $\phi(x)^r$ is itself an appropriate sesquilinear form.

Now considerably more might be asked for in the way of domains, independence of a given free structure, etc., and considerably more is valid. For example, instead of defining the ground state by vector v in **K**, we might use a sufficiently bounded or regular density matrix; this would change nothing essential. Again, it would be desirable to formulate a result which made no reference to the free field, but only assumed, given the (**K**, *W*) and a regular state of this system (or, equivalently, a generating function representing the expectation value of $W(z)$). But this would not change the qualitative situation which is basically: *renormalized powers of the field exist relative to perfectly general (smooth) vacuums.*

This is important because it gives a straightforward, virtually inescapable interpretation of the fundamental nonlinear partial differential equation of the theory—on the assumption that the CCRs hold at all. Of course, it is always conceivable that some generalization of these is all that is applicable to the physical fields, but it seems rather fruitless to attempt to treat the unbounded range of possibilities that would emerge from this. First it seems necessary to explore definitively the structure of mathematical quantum field theory, on the assumption that CCRs do hold.

13.3. Invariant integration in solution manifolds of relativistic equations

There is no assurance that the Dirac–Heisenberg–Pauli quantum field-theoretic formalism is *the* physically correct extension of elementary quantum mechanics to the realm of non-linear relativistic partial differential equations—assuming, indeed, that non-linear quantum field theory is not a specious illusion in the first place. In retrospect, at least, it is clear that Dirac made the formally simplest *Ansatz* that was visible at the time, and that Heisenberg and Pauli extended this in a formally natural way to relativistic systems. But it is quite conceivable that a more sophisticated extension of the elementary Heisenberg (and/or Schrödinger) procedure is really called for.

There are at least two circumstances that support this standpoint. First, the many technical problems and slow fundamental progress even in the two-dimensional scalar field-theoretic case raise the question of whether the attempt at modern mathematical implementation of the DHP formalism is actually physically well motivated; certainly the starting point was more of a formal than an experimental physical character. Second, since elementary quantum mechanics is not at all Lorentz (or even in, for example, the case of the hydrogen atom, Euclidean)-invariant, it seems possible that some modification in the rather amorphous correspondence principle might be appropriate for the treatment of a fully invariant system. In principle at least, the canonical quantization procedure suggests itself as an alternative.

Here we work with the solution manifold M of the *classical* partial differential equation

$$\Box \phi = m^2 \phi + p'(\phi),$$

which is now known to have cogent existence and to be suitably related to free in- and out-fields in a variety of interesting circumstances. The corresponding canonical field operators can be represented by differential operators in this solution manifold. The resulting field operators satisfy the quantized differential equation within counter terms, in a sense similar to that in the quantized distribution formalism just discussed. Thus, although the formalism begins with a symbolism totally different from that involved in the DHP procedure, in the end it resembles it, and indeed, when the interaction Hamiltonian is at most quadratic, can be seen to be mathematically effectively equivalent to it. It seems fair to say that, apart from the quantitative success of the (partially *ad hoc*) Feynman–Schwinger–Tomonaga version of the DHP theory in quantum electrodynamics, the canonical quantization procedure has virtually comparable credentials for consideration as an approach to a true physical field theory.

In order to obtain a field-theoretic Hilbert space in the canonical quantization theory, a Lorentz-invariant weak measure in the solution manifold M is needed that will be analogous to the isotropic Gaussian measure in the free solution manifold; the interacting measure should relate to the physical vacuum as the Gaussian measure relates to the bare vacuum. This measure is naturally sought from the Lorentz-invariant Riemannian structure in the solution manifold. It is interesting that something of the same difference between the two- and higher-dimensional space–time cases which is found in the quantized distribution approach appears here also; but the gap between the lower- and higher-dimensional cases has a much more potentially bridgeable appearance.

It is known that an invertible linear transformation T on a real Hilbert space is absolutely continuous with respect to the isonormal measure in the space iff $T^*T - I$ is a Hilbert–Schmidt (i.e. of square-integrable trace) operator. Now the Lorentz-invariant Riemannian structure in M may be defined as follows. The tangent vectors η at a point ϕ of M correspond in a natural way to solutions of the first-order variational equation

$$\Box \eta = m^2 \eta + p''(\phi) \eta.$$

For each time t, there is a natural map, say W_t, from the tangent space T_ϕ to the space M_0 of free solutions, which carries each η into the free solution with the same Cauchy data at time t. The Riemannian structure is then

$$G_\phi(\eta_1, \eta_2) = \lim_{A\to\infty} \frac{1}{2A} \int_{-A}^{A} \mathrm{Re}\langle W_t\eta_1, W_t\eta_2\rangle \, \mathrm{d}t$$

$$= \tfrac{1}{2}\langle \eta_1^-, \eta_2^-\rangle + \tfrac{1}{2}\langle \eta_1^+, \eta_2^+\rangle,$$

where $\eta_j^\pm$, $j = 1, 2$, are $\lim_{t\to\pm\infty} W_t\eta_j$. The wave operator, mapping M_0 invariably into M, gives a natural correspondence between M_0 and M which permits the Riemannian structure to be transferred to M_0, physically representing the in-field; it then differs from the free linear Riemannian structure defined by the inner product in M_0 by its non-vanishing curvature.

Since M_0 is linear, a tangent vector to M_0 can be naturally identified with a vector in M_0, and G_ϕ can be represented in the form

$$G_\phi(\eta_1, \eta_2) = \mathrm{Re}\langle (I + K)\eta_1^-, \eta_2^-\rangle,$$

where K is a linear operator whose non-vanishing corresponds to the non-vanishing of the interaction. If K is a Hilbert–Schmidt operator, this measure associated with the Riemannian structure has a density relative to the free measure, which can be given rather explicitly, and we have the key feature required for the definition of an interacting measure in M_0. Calculus on the manifolds M and M_0 shows that

$$I + K = (\mathrm{d}S_\phi)' \, \mathrm{d}S_\phi,$$

where S is the scattering transformation, $\mathrm{d}S_\phi$ is its differential at ϕ, and the adjoint is relative to M_0 as a real Hilbert space with respect to the real part of its complex inner product as real inner product. Thus the main question reduces to: Is $(\mathrm{d}S_\phi)' \, \mathrm{d}S_\phi$ a Hilbert–Schmidt operator?

THEOREM 13.1. $(\mathrm{d}S_\phi)' \, \mathrm{d}S_\phi$ *is a Hilbert–Schmidt operator in two space–time dimensions, for a sufficiently high-power interaction, in a neighbourhood of* $\phi \equiv 0$.

Proof. The first step is to rewrite the variational equation as a first-order abstract evolutionary one

$$u' = Au + B(t)u, \qquad A = -A^*$$

in the free-field state–space M_0, in a familiar way, using the known identification of M_0 with the space of Cauchy data, $D(B^{\frac{1}{2}}) \oplus D(B^{-\frac{1}{2}})$, where $B = (m^2 - \Delta)^{\frac{1}{2}}$, and $D(B^a)$ denotes the completion of the domain of B^a with respect to the natural metric $\langle B^a x, B^a y\rangle$, as a real Hilbert space in this metric. To treat the Hilbert–Schmidt question, it is useful to transform to an L_2-type space, in order to equate the Hilbert–Schmidt norm of an integral operator with the L_2 norm of its kernel.

There results the differential equation
$v' = Pv + Q(t)v$:

$$P = \begin{bmatrix} 0 & B \\ -B & 0 \end{bmatrix}, \quad Q(t) = \begin{bmatrix} 0 & 0 \\ -B^{-\frac{1}{2}}p''[\phi(t,x)]B^{-\frac{1}{2}} & 0 \end{bmatrix},$$

where v has two components, each in $L_2(S)$, where S denotes space. Note that $P = -P^*$. Using the method of variation of constants (i.e. essentially, the 'interaction representation'), the S-operator is the $\lim_{t \to \infty} S(t, -\infty)$, where

$$S(t, -\infty) = I + \int_{-\infty}^{t} Q(s) S(s, -\infty) \, ds.$$

Denoting the rth-power trace norm as $\|\cdot\|_r$, it follows that

$$\|S(t, -\infty) - I\|_r \leq \int_{-\infty}^{t} \|Q(s)\|_p \|S(s, -\infty)\|_\infty \, ds.$$

By an elementary inequality it follows in turn that

$$\|S - I\|_p \leq k \exp\left[\int_{-\infty}^{\infty} \|Q(s)\|_p \, ds\right].$$

It suffices therefore to show that $\int \|B^{-\frac{1}{2}} p''[\phi(t,x)] B^{-\frac{1}{2}}\|_2 \, dt$ is a convergent integral. In two-dimensional space–time, for an equation of the form $\Box \phi = m^2 \phi + g\phi^p$, ($p \geq 5$, p odd, $g > 0$), this follows from conservation of energy and the decay estimate $|\phi(t,x)| = O(|t|^{-\frac{1}{2}})$ uniformly in x, if ϕ_{in} is sufficiently smooth and sufficiently close to $\phi \equiv 0$. Thus $dS_\phi - I$ is Hilbert–Schmidt for all sufficiently regular ϕ near $\phi = 0$.

Note that these estimates can be made quite explicit and used to treat the regularity of $dS_\phi - I$ as a function of ϕ to the Hilbert–Schmidt operators, and similar questions of interest in functional-integration-theoretic contexts. We have worked here modulo non-linear scattering theory, and will only remark that the equations

$$\Box \phi = m^2 \phi + g\phi^p \quad (p \text{ odd}, \geq 5 \text{ for } n = 2, 3; \geq 3 \text{ for } n \geq 4)$$

have effective treatments in a neighbourhood of the zero solution (see [4]). It seems probable that the basic results are valid globally on the solution manifold of the equation, but as yet the only one of the cited equations for which this has been shown is that with $n = 4$, $p = 3$ (see [2]).

In more than two space–time dimensions the Hilbert–Schmidt feature certainly fails, unless there are miraculous cancellations, but the operator probably has an rth-power integrable trace, for sufficiently large r (dependent on the dimensionality of the space). In any event, there is no reason to anticipate, especially in high dimension, that the interacting measure in the solution manifold (assuming it exists) will have a density relative to the free measure; it may well be that the relation between the two measures is too singular to be expressed in these terms. It is probable that there exist other Lorentz-invariant weak probability measures in M_0, besides those invariant under all unitary transformations; and some of these may have the property that transformations of the type $I + X$ are absolutely continuous when X has finite rth-power integrable trace, for the higher values of r which apparently apply in higher-dimensional space–times (e.g. $r = 4$ in four space–time dimensions). At least, the proof of the Hilbert–Schmidt criterion makes essential use of the normality of the measure. It should be interesting and would probably be useful to have examples of such measures; or at least to settle the question, whose solution has made no visible progress in the past 15 years, of the existence of Lorentz-invariant weak measures other than mixtures of isonormal distributions.

References

1. HEPP, K. *Theorie de la renormalisation.* Springer–Verlag, Berlin (1969).
2. MORAWETZ, C. S. and STRAUSS, W. A. Decay and scattering of solutions of a non-linear relativistic wave equation. *Commun. pure appl. Math.* **24**, 1–31 (1972).
3. NELSON, E. A quartic interaction in two dimensions. In *Proceedings of the Conference on mathematical theory of elementary particles*, pp. 69–74. M.I.T. Press, Cambridge (1966).
4. SEGAL, I. Dispersion for nonlinear relativistic equations, II. *Ann. Ecole. norm.,* Supl. s4 **1**, 459–97 (1968).
5. SEGAL, I. Local nonlinear functions of quantum fields. In *Functional analysis* (ed. F. Browder), *Proceedings of Conference in honor of M. H. Stone, May 1968*, pp. 188–210. Springer–Verlag, Berlin (1970).
6. SEGAL, I. Construction of nonlinear quantum processes, II. *Inv. Math.* **14**, 211–41 (1971).

14. Definitions and selected applications of Feynman-type integrals

J. TARSKI

14.1. Examples and definitions

HERE we shall give a review of the work done on Feynman-type integrals, including both the rigorous work dealing with the foundations and the applications to specific problems. The rigorous work so far comprises only a few articles, and seems to be a long way from what even the simpler of applications require. Still, the available results can furnish an introduction to problems which may arise.

The integrals we have in mind are in particular those characterized by the weight factor $\exp(iA)$, where A is the action of the physical system. This includes the path-integrals for the (quantum) particles and the history-integrals for fields. There are, besides, some problems which lead to integrals not of this form, e.g. to a Gaussian integral with a complex variance. However, the subsequent discussion can be readily adapted to such Gaussian integrals.

We will present several alternative definitions for Feynman-type integrals. None of the definitions can at present be regarded as superior over the others, and all may produce some useful tools. We ourselves, however, tend to favour the definition based on generalized measures. Before reviewing these definitions, we will give several forms of Feynman-type integrals, and will comment on the history of the subject.

A basic formula for the path-integral is the following (where the $\mathbf{x}_k \in \mathbf{R}^n$, usually $n=3$, the t_k refer to the time, and m below is the mass),

$$G(t_1, \mathbf{x}_1; t_0, \mathbf{x}_0) = N \int_{\eta(t_k)=\mathbf{x}_k} \mathcal{D}(\eta)\, e^{iA(\eta)}. \tag{14.1}$$

Here G is the Green's function for the corresponding Schrödinger equation and A is the classical action, e.g.

$$A(\eta) = \int_{t_0}^{t_1} d\tau \left\{ \tfrac{1}{2} m \sum_{j=1}^{n} [\dot{\eta}^j(\tau)]^2 - V[\eta(\tau)] \right\}. \tag{14.2}$$

The functions η are real, the entity $\mathcal{D}(.)$ is, heuristically, an infinite product of Lebesgue measures, and is invariant under translation,

$$\mathcal{D}(.) = \mathcal{D}(. + \alpha). \tag{14.3}$$

Finally, N in (14.1) is a normalizing factor that will depend on the definition of the integral.

We will write $\mathcal{D}(.)$ quite generally in functional integrals, and two such entities occurring in different integrals need not be related in a simple way. If need be, we can write $\mathcal{D}_1$, $\mathcal{D}_2$, etc.

In (14.1) we may set $x_0 = 0$, $t_0 = 0$, and $m = 1$, and consider more general integrands, so as to have

$$\int_{\eta(0)=0} \mathcal{D}(\eta) \, e^{\frac{1}{2}i\langle\dot{\eta},\dot{\eta}\rangle} f_0(\eta). \tag{14.4}$$

The restriction $\eta(t_1) = x_1$ can come through a factor $\delta[\eta(t_1) - x_1]$ in f_0. A change of the variable function brings (14.4) to the form

$$\int \mathcal{D}(\xi) \, e^{\frac{1}{2}i\langle\xi,\xi\rangle} f(\xi). \tag{14.5}$$

Furthermore, several works recently discussed the following variant of (14.1),

$$G(t_1, \mathbf{x}_1; t_0, \mathbf{x}_0) = N \int_{\mathbf{q}(t_k)=\mathbf{x}_k} \mathcal{D}(\mathbf{q})\mathcal{D}(\mathbf{p}) \, e^{iA(\mathbf{p},\mathbf{q})}, \tag{14.6a}$$

$$A(\mathbf{p},\mathbf{q}) \int_{t_0}^{t_1} d\tau \left(\sum_{j=1}^{n} p^j \dot{q}^j - \left\{ \frac{1}{2m} \sum_{j=1}^{n} (p^j)^2 + V[\mathbf{q}(\tau)] \right\} \right). \tag{14.6b}$$

An integration over $\mathbf{p}$ in (14.6a), carried out in accordance with the usual rules for Gaussian integrals, brings the integral to the form (14.1). The variable $\mathbf{q} = (q^j)$ in (14.6a) is analogous to η in (14.4) or in (14.1), while $\mathbf{p}$ is analogous to ξ in (14.5).

In the case of quantized fields, we have the following formula for time-ordered functions.

$$\langle [F(\phi)]_+ \rangle_0 = \int \mathcal{D}(\eta) \, e^{iA(\eta)} F(\eta), \tag{14.7}$$

where we assumed a single scalar field with action A, and integration is over the histories of the field. The time interval is $(-\infty, \infty)$. The propagator of the theory can be obtained by taking $F(\phi) = \phi(x)\phi(y)$. Moreover, we can give an expression which is intermediate between (14.1) and (14.7), namely, a Green's functional,

$$G(t_1, \zeta_1; t_0, \zeta_0) = N \int_{\eta(t_k,\mathbf{x})=\zeta_k(\mathbf{x})} \mathcal{D}(\eta) \, e^{iA(\eta)}. \tag{14.8}$$

Here the time interval for integration is $[t_0, t_1]$, and N is a normalizing factor.

With regard to the history, it seems fair to say that the path integral had its roots in a paper of Dirac of 1933 [1] who elaborated on the theme that '... it would seem desirable to take up the question of what corresponds in the quantum theory to the Lagrangian method of the classical theory'.

Feynman took up the subject afterwards in his thesis, and in his well-known article of 1948 [2]. Shortly thereafter he applied these ideas in his basic works on quantum electrodynamics [3]. Around 1954 several other authors independently exploited such integrals for fields and constructed solutions, necessarily heuristic, to more general quantized fields with interaction ([5]–[9]). These solutions have the general form of eqn (14.7).

The first attempt at a rigorous construction of the path-integral was apparently that of Cameron in 1960 [10]. He observed that there is no measure (having the usual properties) for the path-integral, even if the exponent in $\exp(iA)$ is modified so as to have a real part. He adapted therefore the approximation scheme which is familiar in case of the Wiener integral (cf. [38]), where the time interval $[0, t]$ is first broken into n parts by the separating points

$$0 = t_0 < t_1 < \ldots < t_n = t. \tag{14.9}$$

The paths η are then assumed to be polygonal, with the successive corners at $(t_j, \eta(t_j))$. The functionals $\exp(iA)$ are evaluated for such polygonal paths, and one integrates over the values $\eta(t_j)$ and passes to the limit $\max(t_j - t_{j-1}) \downarrow 0$. For path-integral, the integrals over $\eta(t_j)$ must be defined with the aid of a convergence factor, e.g. by letting $m \to m + i\delta$, where $\delta \downarrow 0$ subsequently.

Cameron showed that the integrals evaluated as above converge to a path-integral satisfying (14.1), if the potential V satisfies some rather strong conditions, which include in particular some analyticity.

We can point out two drawbacks to this approach. First, the method appears awkward in the case of more intricate systems like fields. Second, the method does not seem to tie in with any convenient mathematical structure, and this would tend to handicap the development of such an integration theory.

Next, there are two methods introduced by Nelson [14]. The first amounts to interpreting the path-integral in terms of Trotter's formula (valid under suitable conditions),

$$\exp[t(A+B)] = \lim_{n \to \infty} [\exp(tA/n) \exp(tB/n)]^n. \tag{14.10}$$

We may take $A = -iH^{(0)}$, where $H^{(0)}$ is the free Hamiltonian, and $B = -iV$. Then the successive approximations in (14.10) form a variant to the successive approximations which occur in Cameron's method, and the limit can be identified with the path-integral.

The second method of Nelson depends on replacing m in the action A by im, so as to have a (measure-theoretic) Wiener integral. One then establishes analyticity in m, and continues from the positive-imaginary to the positive-real values.

172 *Definitions and selected applications*

The first of these methods has the drawback that it does not define the integration of polynomials and of various other functionals. The second is applicable only to situations with extensive analyticity. However, both the analytically continued form of history-integrals, and the Trotter formula, have shown their usefulness in recent work on Euclidean field theory [43].

The last method which we discuss is that suggested by Itô [15]. We consider the path-integral in the form (14.5), and we define the symbol $\mathscr{D}(.)$ as a limit of measures. The form (14.5) presupposes a Hilbert space $\mathscr{H}$. Let $d\mu_{T,\alpha}$ be the Gaussian measure on $\mathscr{H}$ defined by the covariance operator T and the mean vector α (then T must be of trace class, symmetric, and strictly positive). We set

$$I_{T,\alpha}(f) = \frac{1}{c_T} \int d\mu_{T,\alpha}(\xi) \, e^{\frac{1}{2}i\langle\xi,\xi\rangle} f(\xi), \tag{14.11a}$$

where

$$c_T = \int d\mu_{T,0}(\xi) \, e^{\frac{1}{2}i\langle\xi,\xi\rangle}. \tag{14.11b}$$

Then we define

$$\int \mathscr{D}(\xi) \, e^{\frac{1}{2}i\langle\xi,\xi\rangle} f(\xi) = \lim_{T\to\infty} I_{T,\alpha}(f), \tag{14.12}$$

where the limit $T \to \infty$ must be suitably taken and must be independent of α. Itô established the convergence of the integrals in (14.12), and the validity of eqn (14.1) for a very limited class of functionals f with respect to potentials V.

This construction was adapted by the present author to the free scalar field [17]. Here the action can be put into the form

$$iA(\eta) = \tfrac{1}{2}i\langle\eta, B\eta\rangle - \tfrac{1}{2}\langle\eta, C\eta\rangle, \tag{14.13}$$

where B and C are bounded symmetric operators on a suitable Hilbert space. The operator C is chosen so as to provide the increment in $m^2 \to m^2 - i\varepsilon$. Thus $C > 0$, $C \propto \varepsilon$.

It appeared in the above examples that the operators T occur in certain formulas in an awkward way, and that approximating sequences other than $\{d\mu_{T,\alpha}\}$ may be more useful. We were thus led to defining the generalized measures, quite generally, as suitable limits of measures [42].

Clearly, generalized measures can be introduced also for various measure-theoretic integrals. A measure may then factorize into a translationally invariant generalized measure and a weight factor. There are examples where the two factors, i.e. the generalized measure and the weight, have quite different physical interpretations.

But let us return to the path-integral. If a construction is like Itô's, so that we deal with the space of paths as a whole, we will say that we have the global or the Hilbert-space approach. C. DeWitt also contributed to this approach in her construction of a path-integral with the help of the theory of distributions [18].

For contrast, we should like to call an approach based on breaking up the time-interval—the time-evolution approach. The construction of Cameron and the one based on Trotter's formula are of this kind (the analytic continuation in m from the Wiener integral could be classified either way!). Other constructions based on the time-evolution approach can be found in [11]–[13], [16].

14.2. Some problems of integration

We shall comment below on five types of problems. Only in the case of the first one are there rigorous results available for infinite-dimensional integrals.

(1). One basic problem is that of integrability. Examples of integrable functions (or functionals) can of course be found in the cited articles [10]–[18]. For the case of Itô's definition, the results are rather limited, and it appears reasonable to try to extend them by studying the finite-dimensional case. Let us, consider for definiteness

$$I(f) = \frac{1}{(2\pi i)^{n/2}} \lim_{\varepsilon \downarrow 0} \int_{\mathbf{R}^n} d^n u \, e^{\frac{1}{2}(i-\varepsilon)\langle u, u\rangle} f(u). \qquad (14.14)$$

C. DeWitt pointed out that the weight $\exp(\frac{1}{2}i\langle u, u\rangle)$ is in L. Schwartz's spaces $\mathcal{O}_M$ and $\mathcal{O}'_C$, so that $I(f)$ is defined for $f \in \mathcal{O}'_M$ and $\in \mathcal{O}_C$ [18], the procedure with $\lim_{(\varepsilon \downarrow 0)}$ being unnecessary here. The space $\mathcal{O}_C$ is that space of functions whose dual is $\mathcal{O}'_C$. Furthermore, the case $f \in L_1$ is trivial, and the case where f is the Fourier transform of a measure of bounded absolute variation was treated in [15]. Certain other classes of integrable functions are noted in a forthcoming paper [19].

It would also appear desirable to have some more general criteria for integrability, e.g. an adaptation of the bounded convergence theorem. We observe that this theorem cannot hold in the usual form, and it is instructive to look at two counter-examples. For the first, we have the integrable function $f_0(u) = 2$, and the non-integrable one $\exp(-\frac{1}{2}i\langle u, u\rangle)$. The second example is that of a real, non-negative, non-decreasing function f on $\mathbf{R}^1$, which is bounded by the integrable $f_0(u) = u^2$, and which we will define presently.

Consider the real part of $I(f)$, and the zeros of the weight $\cos(\frac{1}{2}u^2)$. They are at $|u|$ equal to

$$a_n = (3\pi + 4n\pi)^{\frac{1}{2}} \quad \text{and} \quad b_n = (\pi + 4n\pi)^{\frac{1}{2}}, \qquad n = 0, 1, 2, \ldots \tag{14.15}$$

At a_n the cosine becomes positive as $|u|$ increases, and at b_n, negative. Let

$$f(u) = \begin{cases} 0 & \text{for } u < a_0 \tag{14.16a} \\ a_n^2 \bullet & \text{for } a_n \leq u < a_{n+1}. \tag{14.16b} \end{cases}$$

The following estimate is adequate for us,

$$C_n =_{df} \int_{a_n}^{a_{n+1}} du \cos(\tfrac{1}{2}u^2) = a_n \left[\int_0^{(b_{n+1}/a_n)-1} dv + \int_{(b_{n+1}/a_n)-1}^{(a_{n+1}/a_n)-1} dv \right] \times$$
$$\times \sin\left[\tfrac{1}{2}a_n^2(2v + v^2)\right] \geq (\text{constant})n^{-\frac{3}{2}} > 0, \tag{14.17}$$

for sufficiently large n. (In the first integral with respect to v the sine is positive, and in the second negative. If n is large then v remains small, and an expansion is effective.) Hence $\sum a_n^2 c_n$ diverges. The divergence as $\varepsilon \downarrow 0$ in (14.14) can now be easily established, and also follows from the results of [19].

It is clear that the function f can be modified to a $\mathscr{C}^\infty$, strictly increasing function, without changing the conclusion.

(2). In the case of infinite-dimensional integrals, a basic problem is that of the space of integration: On what space must f be defined, for the integral $I(f)$ to be meaningful? Alternatively, one may ask: What topology is to be imposed on the functions, when they are defined on a Hilbert space?

Of course, only experience can teach us what choices are suitable here. We should therefore like to present two illuminating examples. These examples in fact emphasize the close correspondence between the positive-definite Gaussian integrals and the Feynman-type Gaussian integrals, even though there is no countably additive measure associated with the latter.

The first example is the functional form of the formula $[p, q] = i^{-1}$,

$$\lim_{\varepsilon \downarrow 0} \int \mathscr{D}(\eta) \exp(\tfrac{1}{2}i\langle \dot\eta, \dot\eta\rangle) \exp\left\{ -i \int d\tau' V[\eta(\tau')] \right\} \times$$
$$\times [\dot\eta(\tau+\varepsilon)\eta(\tau) - \eta(\tau)\dot\eta(\tau-\varepsilon) - i^{-1}] = 0. \tag{14.18}$$

See [27] for heuristic derivation. (We assumed here a one-dimensional system with a continuous potential V and mass unity.) This equation harmonizes with the choice of the Hilbert space $\mathscr{H}$ determined by the norm $\langle \dot\eta, \dot\eta\rangle$ as the space of integration. Indeed, the elements of $\mathscr{H}$ have derivatives which are measureable but in general not continuous, while eqn (14.18) expresses a mean value (in some sense) of the discontinuity of the derivative in the sample paths.

On the other hand, Gaussian integrals lead to familiar integrability conditions, which require that various functionals be determined by Hilbert–Schmidt or by trace-class operators. We expect such conditions to arise also in the case of the Feynman-type integrals. We find, explicitly, for a symmetric operator B with discrete spectrum and eigenvalues $\lambda_j \geq 0$,

$$\int \mathcal{D}(\eta) \, e^{\frac{1}{2}i\langle\dot\eta,\dot\eta\rangle} \, e^{\frac{1}{2}i\langle B\dot\eta,B\dot\eta\rangle} = \left[\prod_j (1+\lambda_j^2)\right]^{-\frac{1}{2}}. \tag{14.19}$$

The product converges iff B is Hilbert–Schmidt. This amounts to saying that the functional $\exp(\frac{1}{2}i\langle b. .\rangle)$ must be extendible to a larger space, of the form $C^{-1}\mathcal{H}$ (where C is non-singular and Hilbert–Schmidt), in order to be integrable. This is also equivalent to the condition that the functional $\exp(\frac{1}{2}i\langle b. .\rangle)$ be continuous in the $\mathcal{T}$-topology of L. Gross [39].

(3). The problem of the space of integration appears in a different light when degenerate Lagrangians are considered. To illustrate, let us use the form (14.6) for the path-integral, and let us choose the canonical variables P, Q in such a way that the Hamiltonian vanishes. Then

$$G(t_1, Q_1; t_0, Q_0) = N \int_{Q(t_j)=Q_j} \mathcal{D}(Q)\mathcal{D}(P) \, e^{i\langle P,\dot Q\rangle}. \tag{14.20}$$

Heuristically the P-integration yields $\delta(\dot Q)$, so that, with an appropriate normalization, we have

$$G(t_1, Q_1; t_0, Q_0) = \delta(Q_1 - Q_0). \tag{14.21}$$

This result [24]. which is expected for a theory where $H = 0$, shows an effective reduction of the space of integration.

Degenerate Lagrangians also occur in static models (where $m \to \infty$ and $H^{(0)} = 0$), in some formulations of spin [35], and in gauge theories. Further discussion can be found in [26].

(4). There has been some interest recently in the special problems which are brought about by curvilinear coordinates in Feynman-type integrals (e.g. [20], [25], [32]). We shall confine ourselves to recalling the close correspondence between the path- and the Wiener integral, and the available extension of the latter to Riemannian manifolds [36].

(5). The asymptotic behaviour of Feynman-type integrals was investigated by the saddle-point method, in several recent papers. However, with regard to rigorous results, it appears that only the case of one-dimensional integrals has been fully treated [41]. For the leading term one has Kelvin's formula,

$$\int_a^b ds\ e^{iuf(s)}g(s) \approx e^{iuf(\bar{s})}g(\bar{s})\left[\frac{2\pi}{|uf''(\bar{s})|}\right]^{\frac{1}{2}} e^{\pm i\pi/4}. \qquad (14.22)$$

Here we assume that f has exactly one extremum on (a, b), namely at $\bar{s}$, that $f''(\bar{s})$ is defined and not equal to 0, and that g is continuous at $\bar{s}$. The large parameter u is assumed real. The sign in $\pm i\pi/4$ agrees with $\pm uf''(\bar{s}) > 0$.

14.3. Applications

Historically the most significant application of Feynman-type integrals has been to quantum electrodynamics [3]. There the integrals served to streamline the manipulations and therefore helped to clarify the nature of various difficulties like the divergences. These integrals continue to be exploited in quantum field theory, where they may facilitate in particular the isolation of dynamical components of gauge fields [27] and the investigation of partial sums of the perturbation series (e.g. [21]).

An (apparently) minor application of a basic nature has been to non-relativistic problems involving 'bad' potentials, i.e. those where the sum $H^{(0)} + V$ is not densely defined, so that the Schrödinger equation does not provide an adequate description [13, 14, 16]. An alternative approach to such potentials is by way of quadratic forms [40].

We will now describe three other types of applications in a little more detail (but we make no attempt to provide a complete survey of applications).

1. The asymptotic evaluations depend on adapting the formula (14.22) to function-space integrals. The procedure is typically as follows. In the integrand one separates the factor $\exp[iB(\eta)]$, where B depends linearly on the large parameter under consideration. Then the path $\bar{\eta}$ defined by

$$[\delta B/\delta\eta(\tau)]_{\eta=\bar{\eta}} = 0 \qquad (14.23)$$

is expected to provide the dominant effect. One thus supposes that

$$\int \mathcal{D}(\eta)\ e^{iB(\eta)}C(\eta) \approx e^{iB(\bar{\eta})}C(\bar{\eta}) \int \mathcal{D}(\chi)\ \exp[i\langle(\delta^2 B)_{\bar{\eta}}\chi, \chi\rangle], \qquad (14.24)$$

where the second functional derivative $(\delta^2 B)_{\bar{\eta}}$ at $\bar{\eta}$ is a kernel which defines an operator, and we denote the latter by the same symbol.

The last integral is Gaussian, and so can be done in closed form. Its value can be interpreted as the Jacobian resulting from a change of variable, cf. $|uf''(\bar{s})|^{-\frac{1}{2}} \exp(\pm i\pi/4)$ in (14.22).

A detailed presentation of asymptotic estimates can be found in [31]. A short but more mathematical discussion is in [30]. Here we confine ourselves to a few simple examples.

(a) Take for $\exp(iB)$ the contribution of the kinetic energy, and for C that of the potential energy. We recover the result of ordinary perturbation theory.

(b) In case of a potential gV, with g being the large parameter, we cannot simply interchange the roles of $\exp(iB)$ and C in (a). However, a clever change of variables leads to meaningful results [31].

(c) For the semiclassical approximation we take $B = A$ (the action) and $C = 1$. This approximation was first investigated in [4], and was applied in recent years by Pechukas [33] and others to problems in atomic collisions (also to other problems, cf. [35]).

(d) These methods have been applied to the classical problem of scattering of waves by a random medium [23]. We start here by imbedding the problem of the reduced wave equation into the initial-value problem defined by a Schrödinger-like equation.

2. Sometimes the usefulness of a perturbation expansion can be enhanced by rearranging the terms. A closed expression for the quantity of interest (e.g. for a Green's function) might thus serve as a useful guide for first making and then investigating the rearrangements. We give two examples from many-body theory.

(a) The perturbation expansion of a path-integral depends on expanding $\exp(-i \int d\tau V)$, evaluating the successive integrals, or moments, which we denote by M_j, and summing. We may rearrange this sum by introducing the series of cumulants K_j, where

$$1 + \sum_{j=1}^{\infty} M_j = \exp\left(\sum_{j=1}^{\infty} K_j\right), \qquad (14.24a)$$

so that

$$K_1 = M_1, \qquad K_2 = M_2 - M_1^2, \qquad \text{etc.} \qquad (14.24b)$$

Reference [28] points out some possible advantages of the cumulants over the moments, e.g. for certain problems the path-integral yields the following estimate for an asymptotic region,

$$K_2 \approx K_3 \approx \ldots \approx 0. \qquad (14.25)$$

Moreover, the cumulants appear to give better numerical results (than the moments) for the tail of the density of states in disordered systems.

(b) In another approximation scheme [22] one starts with the usual expression for Green's function in quantum field theory (eqn (14.7))

$$G(x-y) = \int \mathcal{D}(\bar{\eta})\mathcal{D}(\eta) \, e^{iA}\eta(x)\bar{\eta}(y). \qquad (14.26)$$

The action A contains a quadratic part $A^{(0)}$ with a kernel function and an interaction A_1. The scheme depends on expanding $\exp(iA_1)$ and adjusting the kernel in $A^{(0)}$ so that G be determined by $A^{(0)}$ alone.

The resulting expansion has led to some useful results concerning the behaviour of systems near their critical points.

3. The last application is of an abstract character. The problem of possible statistics of given particles has been treated by functional methods in two different ways (which we summarize below). In both of these one integrates over a multiply connected space of functions (cf. also [34]). The basic premises in the two cases are rather different, but in each case one concludes that only Bose and Fermi statistics are possible.

(a) One way to proceed is to take as the configuration space for n identical particles the space $\mathbf{R}^{3n} = (\mathbf{x}_1, \ldots, \mathbf{x}_n)$, with the elimination of the points where two of the vectors $\mathbf{x}_i$ coincide, i.e. where $\mathbf{x}_j = \mathbf{x}_k$ for some $j \neq k$. Then we form the space of equivalence classes, where

$$(\mathbf{x}_1, \ldots, \mathbf{x}_n) \equiv (\mathbf{x}_{\pi(1)}, \ldots, \mathbf{x}_{\pi(n)}) \quad \text{for some } \pi \in S_n. \tag{14.27}$$

The resulting space $\bar{\mathbf{R}}$ has S_n as its covering group.

The path-integral expression for the propagator G now includes contributions G_α from different homotopy classes. One may express G as follows,

$$G(t_1, X_1; t_0, X_0) = e^{i\delta} \sum_{\pi \in S_n} \chi(\pi) G_\pi(t_1, \ldots), \tag{14.28}$$

where $X_k \in \bar{\mathbf{R}}$, $\exp(i\delta)$ is an irrelevant phase factor, and the $\chi(\pi)$ form a one-dimensional unitary representation of S_n. There are only two such representations, giving bosons and fermions.

(b) Another way depends on assuming, for particles with spin, a history-integral over a suitable space of functions with spin [35]. In the case of integral spin, and for a fixed time, such a formulation yields directly commutativity, hence symmetry under exchange. On the other hand, for half-odd-integer spin one is led to integrating over spaces of pairs $\{\eta, -\eta\}$ of functions, which transform as spinors. Such spaces are not simply connected. Then if one formulates the exchange of two particles in a natural way, an easy calculation (or a topological argument [37]) shows the following: the functional $F(.)$ which corresponds to the two particles in question undergoes a change of sign.

Thus, in both (a) and (b) parastatistics are ruled out.

References

(a) The articles on Feynman-type integrals published through 1954, largely of historical interest:

1. DIRAC, P. A. M. *Phys. Z. Sowjetunion* **3**, 64 (1933).
2. FEYNMAN, R. P. Thesis, Princeton University (1942), and *Rev. mod. Phys.* **20**, 267 (1948).
3. FEYNMAN, R. P. *Phys. Rev.* **80**, 440 (1950) and **84**, 108 (1951).

4. MORETTE DEWITT, C. *Phys. Rev.* **81**, 848 (1951).
5. BOGOLUBOV, N. N. *Dokl. Akad. nauk. USSR* **99**, 225 (1954) [cf. also ibid. **82**, 217 (1952)].
6. EDWARDS, S. F. and PEIERLS, R. E. *Proc. R. Soc. A***224**, 24 (1954).
7. FRADKIN, E. S. *Dokl. Akad. nauk. USSR* **98**, 47 (1954).
8. GELFAND, I. M. and MINLOS, R. A. *Dokl. Akad. nauk. USSR* **97**, 209 (1954).
9. SYMANZIK, K. *Z. Naturforsch.* **9a**, 809 (1954).

(b) The rigorous articles on Feynman-type integrals. The articles 10–14 and 16 emphasize the time-evolution approach, and the articles 15 and 17–19, the global or Hilbert-space approach.

10. CAMERON, R. H. *J. Math. Phys.* **39**, 126 (1960) and *J. anal. Math.* **10**, 287 (1962).
11. DALETSKII, YU. L. *Russ. math. Surveys* **17**, 1 (1962).
12. BABBITT, D. G. *J. Math. Phys.* **4**, 36 (1963).
13. FELDMAN, J. *Trans. Am. math. Soc.* **108**, 251 (1963).
14. NELSON, E. *J. Math. Phys.* **5**, 332 (1964).
15. ITÔ, K. in *Proceedings of the 5th Berkeley Symposium on mathematical statistics and probability. Vol. II, part 1, P. 145.* University of California Press, Berkeley (1966).
16. FARIS, W. G. *Pacif. J. Math.* **22**, 47 (1967). [cf. also *J. funct. Anal.* **1**, 93 (1967).]
17. TARSKI, J. *Ann. Inst. H. Poincaré A***15**, 107 (1971) (erratum to follow).
18. DEWITT, C. M. *Commun. math. Phys.* **28**, 47 (1972), and preprint.
19. BUCHHOLZ, D. and TARSKI, J. Paper in preparation.

(c) Selected recent articles on the applications of Feynman-type integrals.

20. ARTHURS, A. M. *Proc. R. Soc. A***318**, 523 (1970).
21. BARBASHOV, B. M. and NESTERENKO, V. V. *Teor. mat. Fiz.* **10**, 196 (1972).
22. BHAGAVAN, M. R. and EDWARDS, S. F. *Proc. phys. Soc.* **90**, 953 (1967); BHAGAVAN, M. R. and LAMBERT, P. *J. Phys. A (Gen. Phys.)* **2**, 1 (1969).
23. CHOW, P.-L. *J. Math. Phys.* **13**, 1224 (1972).
24. DAVIES. H. *Proc. Camb. phil. Soc.* **59**, 147 (1963).
25. DOWKER, J. S. Report at this conference.
26. FADDEEV, L. D. *Teor. mat Fiz.* **1**, 1 (1969); FADDEEV, L. D. and POPOV, V. N. *Phys. Lett.* **25B**, 29 (1967).
27. FEYNMAN, R. P. and HIBBS, A. R. *Quantum mechanics and path integrals.* McGraw–Hill, New York (1965), especially p. 175.
28. JONES, R. and LUKES, T. *Proc. R. Soc. A***309**, 457 (1969).
29. LAIDLAW, M. G. G. and DEWITT, C. M. *Phys. Rev.* **D3**, 1375 (1971).
30. MASLOV, V. P. *Teor. mat. Fiz.* **2**, 30 (1970).
31. MCLAUGHLIN, D. W. *J. Math. Phys.* **13**, 784 (1972).
32. MCLAUGHLIN, D. W. and SCHULMAN, L. S. *J. Math. Phys.* **12**, 2520 (1971).
33. PECHUKAS, P. *Phys. Rev.* **181**, 166, 174 (1969).
34. SCHULMAN, L. S. Chapter 12 of this book.
35. TARSKI, J. Paper in preparation.

(d) Other works referred to in the text.

36. EELLS, J. and ELWORTHY, K. D. In *Problems in non-linear analysis* (ed. G. Prodi). CIME session 1970; p. 67. Edizioni Cremonese, Rome (1971).
37. FINKELSTEIN, D. and RUBINSTEIN, J. *J. Math. Phys.* **9**, 1762 (1968).

38. GELFAND, I. M. and YAGLOM, A. M. *J. Math. Phys.* **1**, 48 (1960).
39. GROSS, L. *Mem. Am. math. Soc.* No. 46 (1963).
40. SIMON, B. *Quantum mechanics for Hamiltonians defined as quadratic forms.* Princeton University Press (1971).
41. SIROVICH, L. *Techniques of asymptotic analysis*, Sec. 24. Springer-Verlag, Heidelberg (1971).
42. TARSKI, J. *Ann. Inst. H. Poincaré* A**17**, 313 (1972).
43. VELO, G. and WIGHTMAN, A. (eds). *Constructive quantum field theory, Lecture notes in physics* No. 25. Springer-Verlag, Heidelberg (1973).

15. Problems in quantum gravity

J. G. TAYLOR

15.1. Introduction

BLACK holes and the problem of gravitational collapse have come much to the fore recently. The difficulties presented by such phenomena, especially of the fate of matter at the end point of collapse, are such as to indicate the need for a modification of the classical theory of gravity. The only known way to avoid the problems of collapse are to include the quantum effects which modify the behaviour of matter in such extreme conditions. In particular, particle annihilation or creation could considerably modify the classical results. However, further difficulties are then presented by the equation

$$G_{\mu\nu} = \kappa T_{\mu\nu}, \qquad (15.1)$$

where $G_{\mu\nu}$ on the left of (15.1) is the usual divergence-free tensor ($R_{\mu\nu} - \frac{1}{2}Rg_{\mu\nu}$), with $R_{\mu\nu}$ and R the curvature tensor and scalar respectively for the metric tensor $g_{\mu\nu}$ and κ the gravitational coupling constant; $T_{\mu\nu}$ is the energy–momentum tensor of matter (including the gravitational contribution). If the right-hand side of (15.1) is a quantum field operator then so should the left-hand side, unless the right-hand side is to be interpreted as some expectation value $\langle T_{\mu\nu} \rangle$ of $T_{\mu\nu}$ in some state $>$. There is no indication of how the state $>$ is to be defined, so that the resulting theory is not satisfactory. We will accept that (15.1) is to be extended so that if the right-hand side is a field operator then so is the left-hand side. Thus we consider the problem of the quantization of the gravitational field itself.

The Lagrangian for this theory can be written, for the gravitational field alone, as

$$L_E = R\sqrt{(-g)}, \qquad (15.2)$$

where $g = \det g_{\mu\nu}$. With the usual definitions

$$g^{\mu\nu}g_{\nu\lambda} = \delta^{\mu}_{\lambda},$$

$$\Gamma^{\mu}_{\nu\lambda} = \tfrac{1}{2}g^{\mu\sigma}(\partial_\nu g_{\sigma\lambda} + \partial_\lambda g_{\sigma\nu} - \partial_\sigma g_{\nu\lambda}), \qquad (15.3)$$

$$R_{\mu\nu} = \partial_\nu \Gamma^{\sigma}_{\mu\sigma} - \partial_\sigma \Gamma^{\sigma}_{\mu\nu} + \Gamma^{\rho}_{\mu\sigma}\Gamma^{\sigma}_{\nu\rho} - \Gamma^{\rho}_{\mu\nu}\Gamma^{\sigma}_{\rho\sigma},$$

$$R = R^{\mu}_{\mu},$$

we see that the Lagrangian (15.2) is quadratic in the first derivative of $g_{\mu\nu}$, but involves the field variable $g_{\mu\nu}$ itself in a non-polynomial fashion. The analogous single-variable Lagrangian, with the variable denoted by $q(t)$, is

$$L = f(q)\dot{q}^2, \qquad (15.4)$$

where $f(q)$ can only be expressed by a non-finite series in q.

Such a non-polynomial Lagrangian has a number of problems associated with it when one tries to quantize it.

(1) The ordering of the non-commuting variables $q, \dot{q}$ in (15.4) is ambiguous.

There are many quantum-mechanical Lagrangians corresponding to the classical Lagrangian of (15.4).

(2) The interaction term in the Lagrangian (15.4),

$$L_{\text{int}} = (f-1)\dot{q}^2,$$

also involves the velocity $\dot{q}$ quadratically. Perturbation theory rules may have to be modified due to this.

(3) There are expected to be a large number of ultraviolet divergencies on evaluation of the field-theoretic version (15.2) of (15.4); the theory will be highly non-renormalizable.

(4) There are various prescriptions to avoid these divergences, but it is necessary to be careful that the symmetry of the theory—coordinate invariance in our case—be preserved when such rules are used.

(5) It is not even clear that the results of calculations are independent of the choice of dynamical variable to quantize, such as $g_{\mu\nu}$ or $g^{\mu\nu}$.

We will turn to discuss these various problems briefly.

15.2. Divergences and ambiguities

The ultraviolet divergencies arise from divergent integrals. These can be handled by the method of dimensional regularization, where the archetypal integral $\int d^4q [q^2(p-q)^2]^{-1}$ is replaced by the function $I(\omega)$ of the complex variable ω,

$$I(\omega) = \int d^{2\omega} q [q^2(p-q)^2]^{-1}$$
$$= \pi^\omega (p^2)^{\omega-2} \Gamma(2-\omega) B(\omega-1, \omega-1). \qquad (15.5)$$

It is usual to evaluate (15.5) at the pole at $\omega = 2$ *contained in* $\Gamma(2-\omega)$ by subtracting the pure pole contribution and cancelling it. However, this procedure is ambiguous since we could have started with $f(\omega)I(\omega)$ instead of $I(\omega)$, for any function f with $f(2) = 1$. Then the remainder in (15.5) at $\omega = 2$ will have an additional term involving $f'(2)$, so will contain an arbitrary constant. Higher-order terms involve higher poles of ω at 2, and so introduce further arbitrary constants. This is to be expected in such a non-renormalizable theory; there appears presently to be no way of avoiding this degree of ambiguity without making *ad hoc* hypothesis. This is certainly true when superpropagator methods are used; such a transcendental function as $\exp[\lambda^2 \Delta(x)]$ contains an infinite number of arbitrary constants in its definition.

Owing to this non-renormalizability it is necessary to investigate *all* the arbitrary constants in this theory.

15.3. Choice of variable

There are many alternate choices of variable to quantize:

$$g_{\mu\nu},\ g^{\mu\nu},\ g^{\beta}g^{\mu\nu},\ g^{\beta}g_{\mu\nu} = \chi_{\mu\nu},\ \text{etc.}$$

If we use $\chi_{\mu\nu}$ then it is possible to express the Lagrangian (15.2) as

$$L_E = \tfrac{1}{2}(-\chi)^{-2-[(1+6\beta)/2(1+4\beta)]} P_q^{\lambda\mu\nu,\alpha\beta\gamma} \partial_\lambda \chi_{\mu\nu} \partial_\alpha \chi_{\beta\gamma}, \tag{15.6}$$

where $P_q^{\cdots}$ is a polynomial in χ of degree q. We note that L_E becomes purely polynomial in the case that

$$\beta = -(5+2m)/2(1+4m), \tag{15.7}$$

where $m = 0, 1, 2, \ldots$; for such values of m, $\beta = -\tfrac{5}{22}, -\tfrac{7}{30}, \ldots$. Then

$$L_E = \tfrac{1}{2}\chi^m P_q^{\lambda\mu\nu,\alpha\beta\gamma} \partial_\lambda \chi_{\mu\nu} \partial_\alpha \chi_{\beta\gamma}, \tag{15.8}$$

so that L_E is at least of degree 11 in χ (when $m = 0$). Thus even though L_E can be made polynomial it does not become renormalizable, for which we need a constant multiplying the quadratic derivative expression $\partial_\lambda \chi_{\mu\nu} \partial_\alpha \chi_{\beta\gamma}$. In any case when matter is present the factor $\sqrt{(-g)}$ contributes the factor $\chi^{-(1+4m)/18}$, which can never be polynomial for positive integral m.

15.4. Operator ordering

The various quantum-mechanical Lagrangians derived from (15.2) by choice of different operator orderings are also required to be invariant under the coordinate transformation

$$x \to x - \xi(x),$$
$$g_{\mu\nu} \to g_{\mu\nu} - \xi^\gamma \partial_\gamma g_{\mu\nu} - g_{\mu\gamma} \partial_\nu \xi^\gamma - g_{\nu\gamma} \partial_\mu \xi^\gamma, \tag{15.9}$$
$$g^{\mu\nu} \to g^{\mu\nu} - \xi^\gamma \partial_\gamma g^{\mu\nu} + g^{\mu\gamma} \partial_\gamma \xi^\nu + g^{\nu\gamma} \partial_\gamma \xi^\mu.$$

It is possible that this condition is satisfied by only a small number of these Lagrangians; it is to be hoped only one. This is actually the case in another non-polynomial quadratically-derivative interaction theory, that of chiral $SU(n) \times SU(n)$.

In that case the Lagrangian is basically

$$L = \tfrac{1}{2} g_{ij}(\phi) \partial_\mu \phi^i \partial^\mu \phi^j, \tag{15.10}$$

being invariant under the field transformations

$$\delta \phi^i = \xi_a^i(\phi) \theta^a$$

for infinitesimal θs, the ξs being the Killing vectors for (15.10), so that

$$g^{ij} = \xi_a^i \xi_a^j.$$

There is then a unique quantum-mechanical Hamiltonian for (5.10) consistent with the symmetry group, as has been shown recently by Charap,

$$H = \tfrac{1}{8}[\pi_i,[\pi_j, g^{ij}]_+]_+ + \tfrac{1}{2}\nabla\phi^i \cdot \nabla\phi^j \cdot g_{ij} + \tfrac{1}{8}[\delta^3(0)]^2 \xi^i_{a,j}\xi^j_{a,i}, \quad (15.11)$$

where the momenta π_i are defined from the fields and their derivatives by

$$\pi_i = \tfrac{1}{2}[g_{ij}, \dot{\phi}^j]_+.$$

It is possible to obtain perturbation rules from (15.11) by following the steps:

Hamiltonian + canonical commutation relations (CCRs)
→ field equations + CCRs.
→ Green's functional equations
→ functional integral solution
→ perturbation rules.

The detailed working out of these steps, along with references to earlier work, is given explicitly in [1], and so it will not be discussed here except to point out the final cancellation of the $[\delta^3(0)]^2$ term in (15.11) does occur, at least to the lowest non-trivial order. There is also a $\delta^4(0)$ term which can also be shown to cancel, but not exactly; it may be avoided *ab initio* by a suitable time-averaging definition necessary to obtain the Green's functional equations [1].

It would be nice to report that a similar discussion has been given for gravity. Unfortunately this is not the case. There does not appear to exist any ordering ensuring coordinate invariance for the choice of $g^{\mu\nu}$, $g_{\mu\nu}$, $g^\beta g^{\mu\nu}$, $\chi_{\mu\nu}$, or L^μ in the Weyl or Møller forms of the Lagrangian. Beyond that is the more difficult point that the infinitesimal transformation (15.8) is itself not even well defined owing to the lack of commutation of the variables $\xi^\gamma(g)$ and the derivatives $\partial_\alpha g^{\mu\nu}$ or of $\partial_\alpha \xi^\gamma$ with $g^{\mu\nu}$. Thus one has also to attempt to vary the ordering in (15.8) to achieve invariance. This has not proved possible, and indeed may not be so. However, we can settle for much less than this. For if we can obtain manifestly generally covariant expressions for observables then we should be satisfied if at the same time these expressions were unitary, divergence-free, and satisfied all the other usual conditions of quantum field theory (causality, positive-definiteness, mass-spectral conditions, etc.). We turn to that now.

15.5. The loop expansion method

Earlier attempts to evaluate expressions in quantum gravity have involved the expansion of $g_{\mu\nu}$ around the Lorentz metric $\eta_{\mu\nu}$,

$$g_{\mu\nu} = \eta_{\mu\nu} + \kappa h_{\mu\nu}. \quad (15.12)$$

Problems in quantum gravity

The resulting perturbation expansion in powers of κ is sadly not generally covariant term by term, so that after regularization further conditions have to be imposed on the finite remainder to preserve the coordinate invariance. It is clear that these extra restrictions still allow for an infinite number of arbitrary parameters in the perturbation expansion, so that what proves to be the very considerable task of determining these conditions does not improve the situation.

A more recent approach has allowed the general covariance of the theory to be preserved from the beginning. It uses an expansion about a classical background field in ever-increasing numbers of closed loops of intermediate particles; each term in this expansion may be shown to be coordinate invariant. To achieve this we use the formal functional integral for the Green's functional $G(J)$, the generating function of the infinite set of Green's functions $\langle 0|T[A(x_1)-A(x_n)]|0\rangle$ for a quantum field $A(x)$,

$$G(J) = \int \prod_x dA(x)\, e^{iW(A)+JA}, \qquad (15.13)$$

where $W(A)$ is the classical action for A derived from the Lagrangian $L(A)$,

$$W(A) = \int L(A)\, d^4x$$

and

$$JA = \int J(x)A(x)\, d^4x.$$

We write

$$A = A_{cl} + \psi,$$

where A_{cl} is defined so as to satisfy the classical field equation in the presence of the source J

$$W'(A_{cl}) = iJ. \qquad (15.14)$$

Then

$$G(J) = \exp\{iW(A_{cl}) + JA_{cl}\} \int \prod_x d\psi(x)\, \exp\{i[\tfrac{1}{2}\psi^2 W'' + O(\psi^3)]\}. \qquad (15.15)$$

If we consider the modified Green's functional

$$G_\lambda(\lambda^{-1}J) = \int \prod_x dA(x)\, \exp\left\{\frac{i}{\lambda}W(A) + \frac{1}{\lambda}JA\right\} \qquad (15.16)$$

the expansion of (15.16) in powers of λ^{-2} can be shown to provide the loop expansion. Working only with connected parts, with the suffix c, it can be shown explicitly that

$$G^c_\lambda(\lambda^{-1}J) = \ln G_\lambda(\lambda^{-1}J) = \lambda^{-2}G^c_{-1}(J) + G^c_0(J) + \lambda^2 G^c_1(J) + \ldots + \lambda^n G^c_n(J), \\ + \ldots, \qquad (15.17)$$

where $G^c_n(J)$ is the generating functional of the sum of all perturbation diagrams with arbitrary numbers of external lines but exactly $(n+1)$ internal closed loops. We can evaluate from (15.16) and (15.17) that

$$A_{cl} = \delta G^c_{-}/\delta J(x), \qquad (15.18)$$

$$-\tfrac{1}{2}\mathrm{tr}\ln W''(A_{cl}) = G^c_0(J), \qquad (15.19)$$

$$G_1 = \left[-\frac{1}{2}\left(\frac{1}{3!}\right)^2 \frac{1}{2} W_{,lmn} W_{,pqr}(G_{lm}G_{np}G_{qr} + \text{permutations}) + \right.$$

$$\left. + \frac{i}{4!} W_{,lmnp}(G_{lm}G_{np} + \text{permutations}) \right] \frac{(i\pi)^{\frac{1}{2}n}}{4} \cdot (\det W_{,ab})^{-\frac{1}{2}}, \qquad (15.20)$$

where

$$G_{lm} = \delta G_{-1}/\delta J(x_l)\delta J(x_m), \quad W_{,ij\ldots} = \delta W/\delta J(x_i)\delta J(x_j)\ldots$$

It is possible to write down the higher loop expressions, but these are clearly very complicated so we will not give them here.

15.6. Counter terms in the loop expansion

The crucial question we now have to answer is what the form of counter terms is necessary to remove any ultraviolet divergencies arising from the internal particle loops. This can be solved for a scalar field by expansion of L about the classical field value A_{cl} [2]. If, for small ψ,

$$L(A_{cl} + \psi) = \partial_l\psi\partial^\mu\psi + \psi M(A_{cl})\psi + O(\psi^3), \qquad (15.21)$$

then the single-loop generating functional has divergences all contained in $-\tfrac{1}{2}\mathrm{tr}\ln(1 - \Box^{-1}M)$. The divergence in this arises purely from the term in the logarithm quadratic in M, and can be removed in the dimensional-regularization scheme by the counter term

$$\Delta_1 L = (\omega - 2)^{-1} M^2(A_{cl}). \qquad (15.22)$$

For the Einstein Lagrangian (15.2) we obtain [3]

$$\Delta_1 L = \frac{\sqrt{(-g)}}{\omega - 2}\left(-\frac{22}{15}R^2 + \frac{49}{10}R_{\alpha\beta}R^{\alpha\beta}\right).$$

On the mass shell, when $R = R_{\mu\nu} = 0$ then $\Delta L = 0$ and the theory is convergent. When matter is present ([3], [4], [5])

$$\Delta_1 L = \frac{\sqrt{(-g)}}{\omega - 2}\left[-\frac{211}{144}R^2 + \frac{589}{120}R_{\alpha\beta}R^{\alpha\beta} - \frac{19}{12}R\partial_\mu\phi\partial^\mu\phi + \frac{3}{2}R_{\mu\nu}\partial^\mu\phi\partial^\nu\phi + \right.$$
$$\left. + \frac{9}{16}(\partial_\mu\phi\partial^\mu\phi)^2 + \frac{1}{2}(\partial_\mu\partial^\mu\phi)^2\right]. \tag{15.23}$$

Now on mass shell, when $D^2\phi = 0$, $R = -\frac{1}{2}\partial_\mu\phi\partial^\mu\phi$,

$$\Delta_1 L = \frac{\sqrt{(-g)}}{\omega - 2} \cdot \frac{6377}{1440} R^2 \neq 0.$$

However, addition of (15.23) to the original Lagrangian produces further simple-loop divergences, and so further single-loop counter terms; these are rapidly seen [5] to lead to higher divergences with increasing powers of $\partial_\mu\phi\partial^\mu\phi$, etc. The possibility of cancellation between these terms, even on mass shell, seems very remote, so that this approach is soon beset with difficulties.

On the other hand, one can use the quartic derivative of $g_{\mu\nu}$ in $\Delta_1 L$ to obtain a propagator for the graviton (and all other particles) decreasing at infinite (energy)^2 p^2 as $(p^2)^{-2}$. This ensures that no new divergencies are present, so that no new counter terms will be necessary. However, the heavy price of the presence of ghosts has to be paid for then, so it is a very mixed blessing.

15.7. Conclusion

There is at present no consistent quantum field theory of gravity. There are numerous difficulties in the way of achieving such a theory but the most crucial is that of ultraviolet divergences and the associated ambiguities. There is no satisfactory solution to this problem. The only way to proceed is to modify Einstein's theory so as to incorporate suitable fields giving further cancellations. Such possibilities are presently being explored. It is to be hoped that they allow a unification of gravity with the other forces of nature.

References

1. TAYLOR, J. G. *J. Phys. A* **1**, 12 (1974).
2. T'HOOFT, G. *Nucl. Phys. B* **62**, 444 (1973).
3. T'HOOFT, G. and VELTMAN, T. One loop divergences in the theory of gravitation. 1723, CERN preprint (1973).
4. DESER, S. and NIEUWENHURZEN, P. VON, One loop divergences of quantised Einstein–Maxwell fields. Brandeis preprint (1974).
5. TAYLOR, J. G. and NOURI, M. The non-renormalisibility of gravity. King's College, London preprint (1974).

Author Index

Amit, D. J. 129, 135
Ampère, A. M. 58
Anderson, J. L. 36
Anderson, P. W. 127, 134
Arthurs, A. M. 179

Babbitt, D. G. 179
Babich, V. M. 156
Balian, R. 47, 156
Barbashov, B. M. 179
Bardeen, J. 135
Bargmann, V. 81
Baxendale, P. 65
Berger, M. 50
Bergmann, P. G. 50
Berry, M. V. 156
Bessaga, C. 66
Bhagavan, M. R. 179
Bleistein, N. 156
Bloch, G. 47, 156
Bochner, S. 50
Bogolubov, N. N. 179
Bonnesen, T. 14
Born, M. 43, 50
Brockett, R. W. 67
Browder, F. 168
Buchholz, D. 179
Burghelea, D. 67
Buslaev, V. S. 96

Cameron, R. H. 33, 171, 179
Cheng, K. S. 36, 43, 50
Chernoff, P. R. 50
Chow, P.-L. 179
Clark, J. M. C. 67
Cohen, L. 50
Cooper, L. N. 135
Coyne, S. 156

Daletskii, Yu. L. 60, 63, 65, 67, 179
Danilov, Yu. P. 156
Davies, H. 179
Deam, R. 59
Debiard, A. 67
De Broglie, L. 50
De Gennes, P. G. 59
De Haas-Lorentz, 43
Denton, R. 135
Deser, S. 187
De Witt, B. S. 35, 36, 37, 47, 146, 156
De Witt, C. M. 51, 147, 156, 173, 179
Dirac, P. A. M. 35, 170, 178
Dobrushin, 69
Doob, J. L. 98, 100, 122
Dowker, J. S. 51, 147, 156, 179
Dudley, R. M. 14, 67
Dynkin, D. 51

Ebin, D. G. 67
Edwards, S. F. 59, 124, 134, 146, 156, 179
Eells, J. 60, 65, 67, 179
Einstein, A. 97, 122
Elworthy, K. D. 67, 179
Erdelyi, A. 153, 156
Eskin, L. D. 51
Everts, U. 126, 134

Faddeev, L. D. 96, 179
Faris, W. G. 179
Feldman, J. 14, 67, 179
Fenchel, W. 14
Feynman, R. P. 35, 36, 51, 126, 134, 171, 178, 179
Finkelstein, D. 179
Fischer, A. E. 67
Flaschka, H. 96
Folias, C. 82, 106, 123
Ford, G. W. 99, 107, 122
Fortuin, C. M. 14
Fradkin, E. S. 179
Freed, K. 54
Fröhlich, J. 69, 81
Furry, W. H. 51

Gangolli, R. 51
Gardner, C. S. 96
Gauss, C. F. 58
Gaveau, B. 67
Gelfand, I. M. 51, 142, 156, 179, 180
Gilbert, R. P. 156
Ginibre, J. 14
Girsanov, I. V. 67
Glimm, J. 69, 81, 136, 137, 143
Goodman, V. 67
Grant, J. W. V. 59
Greene, J. M. 96
Griffiths, R. B. 14, 69
Gross, L. 14, 67, 69, 175, 180
Guerra, F. 69, 81
Gulyaev, Y. V. 146, 156

Hamann, D. R. 130, 131, 135
Hamilton, J. F. 51
Hammersley, J. M. 86
Hassing, R. F. 135
Helgason, S. 51
Hénon, M. 96
Hepp, K. 159
Hibbs, A. R. 51, 134, 179
Hubbard, J. 125, 134

Ito, K. 33, 36, 43, 51, 172, 179
Ivoanov, E. N. 50, 52

Jaffe, A. 69, 81, 136, 137, 143
Jensen, H. 51
Jones, R. 179
Jost, R. 143

Kac, M. 15, 33, 96, 99, 107, 122
Kasteleyn, P. W. 14
Keiter, H. 129, 135
Kelly, D. 14
Kim, S. 122
Klander, J. R. 51
Kolmogorov, A. N. 37, 51, 101
Koppe, H. 51
Kruskal, M. D. 96
Kuiper, N. H. 67
Kunz, H. 14
Kuo, H.-H. 60, 61, 63, 67

Laidlaw, M. G. G. 147, 156, 179
Lamb, H. 113, 122
Lambert, P. 179
Langevin, P. 97, 122
Lax, P. D. 106, 120, 122
Le Cam, L. 14, 67
Lévy, P. 51
Lewis, J. T. 122
Lichnerowicz, A. 51
Loomis, L. H. 67
Ludwig, G. 43, 50, 156
Lukes, T. 179

Mandelstam, S. 37, 51
Marsden, J. E. 67
Martin, W. T. 33, 134
Maslov, V. P. 179
Mayes, I. W. 51
Mayne, D. Q. 67
Mazet, E. 67
Mazur, P. 99, 107, 122
McKean, H. P. 33, 43, 47, 63, 67
McLaughlin, D. W. 40, 51, 156, 179
McShane, E. J. 62, 63, 67, 68
Melnikoff, A. 91, 96
Merzbacher, E. 14
Milgram, A. N. 47, 51
Milnor, J. 156
Minakshisundaram, S. 47, 51
Minlos, R. A. 69, 179
Minna, R. M. 96
Moerbeke, P. van 96
Monroe, J. L. 127, 134
Montroll, E. W. 156
Morawetz, C. S. 168
Morette, C. 35, 51, 179
Mount, K. E. 156
Mühlschlegel, B. 134, 135

Nelson, E. 68, 69, 77, 80, 81, 122, 159, 168, 171, 179
Nesterenko, V. V. 179
Newton, R. G. 156
Nieuwenhurzen, P. von 187
Norcliffe, A. 51, 156
Nouri, M. 87

Omori, H. 68
Ornstein, L. S. 97, 101, 105, 123
Osterwalder, K. 69, 76, 82
Ozkaynak, A. H. 77, 82

Patodi, V. K. 51
Pauli, W. 35, 43, 51
Pechukas, P. 177, 179
Peierls, R. E. 179
Penrose, R. 51
Percival, I. C. 156
Perrin, F. 50, 51
Perrin, J. 97, 122
Phillips, R. S. 106, 120, 122
Pietch, M. 63, 67
Pleijel, A. 51
Podolsky, B. 51
Popov, V. N. 179
Postnikov, M. M. 156
Prodi, G. 179
Pulé, J. V. 122

Ramer, R. 60, 65, 68
Rayleigh, Lord, 125
Riesz, M. 46, 51
Roberts, M. J. 156
Rosen, L. 69, 81
Rosenbloom, P. C. 47, 51
Rubinstein, J. 179
Ruse, H. S. 47, 51

Scalapino, D. J. 135
Schiff, L. I. 51
Schilder, M. 151, 156
Schlesinger, L. 51
Schoenberg, I. J. 14
Schouten, J. A. 51
Schrader, R. 69, 76, 82
Schrieffer, J. R. 130, 131, 135
Schrödinger, E. 52
Schulman, L. S. 40, 51, 52, 156, 179
Schwinger, J. 76, 80, 82
Segal, I. E. 68, 82, 134, 168
Sherington, D. 134
Sherman, S. 14
Shnaiderman, Ya. I. 60, 63, 65, 67
Siegert, A. J. F. 124, 126, 127, 134
Simon, B. 69, 81, 180
Singer, I. M. 47, 51

Author Index

Sirovich, L. 180
Smoluchowski, M. von 97, 122
Sommerfeld, A. 52
Spencer, T. 69, 81, 136
Sternberg, S. 67
Stratonovich, R. L. 48, 52, 125, 134
Strauss, W. A. 168
Symanzik, K. W. 82, 143, 179
Sz.-Nagy, B. 82, 106, 123

Tarski, J. 179, 180
Taylor, J. G. 187
Testa, F. J. 52
Thom, R. 155, 156
Thomas, L. C. 122, 123
t'Hooft, G. 187

Uhlenbeck, G. E. 97, 101, 105, 123
Uhlenbrock, D. A. 76

Valiev, K. A. 50, 52
Van Vleck, J. H. 35, 52

Varadhan, S. R. S. 151, 156
Veblen, O. 42, 52
Velo, G. 81, 143, 180
Veltman, T. 187
Vilenkin, N. Ya. 142

Walker, A. G. 47, 52
Wang, M. C. 123
Welker, H. 52
Wiegel, F. W. 134
Wightman, A. 81, 143, 180
Wigner, E. P. 8, 14, 81
Wilkins, J. W. 135
Wilson, K. G. 156
Wolff, P. R. 135

Yaglom, A. M. 51, 156, 180
Yamakawa, H. 54, 59
Yao, T. H. 76, 81
Yosida, K. 52

Zittartz, J. 126, 134

Subject Index

action, 150
Airy function, 152
Anderson model, 124
atomic collisions, 177

Banach manifolds, 60
Bardeen–Cooper–Schrieffer (BCS) interaction, 131
Boltzmann factor, 3
Bose statistics, 178
bosons, 178
Brownian
  motion, 18
  paths, 18
Brownian-motion
  expectations, 16
  measure, 17
  process, 15
Brunn–Minkowski theorem, 2

calculus of variations, 85
Cameron–Martin formula, 24
canonical
  commutation relation, 34
  quantum process, 160
Casimir operator, 43
catastrophes, 155
Cauchy data, 158
caustics, 144
C*-dynamics, 159
chain, 53
chemical potential, 124
chiral dynamics, 35
classical
  mechanics, 94
    optics, 148
    trajectory, 149
computing machinery, 83
conjugate points, 150
conjugation, 75
convolution, 2
correlation function, 57
Coulomb potential, 9
covariant derivative, 34
critical points, 178
crystal lattice, 84
crystalline state, 11
cubical topology, 84
cumulants, 177
Curie law, 130
current densities, 80
curvature tensor, 181
curved spaces, 35
curvilinear coordinates, 175
cylinder set measures, 66

dielectric function, 126
diffusion equation, 7
dilating semigroups of contractions, 105
Dirichlet boundary condition, 137
disordered systems, 57

Einstein Lagrangian, 186
Einstein–Smoluchowski theory, 97
electromagnetic
  caustics, 155
  field, 69
  interaction, 36
electron optics, 155
elliptic partial differential equations, 137
energy gap, 131
energy–momentum tensor, 181
entropy, 55
ergodic theorem, 18
Euclidean
  electromagnetic field, 69
  group, 69
  Proca field, 69
  quantum field theory, 69
Euler–Lagrange equation, 149
equilibrium statistics, 124
equipartition of energy, 98

Feller's test, 24
Fermi
  energy, 128
  statistics, 178
fermions, 11
Feynman
  diagrams, 124
  postulate, 37
  quantization, 35
Feynman–Kac formula, 19
Feyman-type integrals, 169
Fisk–Stratonovich integral, 63
focal points, 150
Fock space, 69
Ford–Kac–Mazur model, 99
Fourier transform, 12
Fubini's theorem, 2
functional integrals, 1

gauge-transformation, 38
Gaussian
  integrals, 57
  measures, 1
  process, 75
  weight, 5
geodesic distance, 40
Gibbs distribution function, 9
Gorkov interaction, 126

Subject Index

gravitation, 35
gravitational
  collapse, 181
  coupling constant, 181
graviton, 187
Green's
  function, 12
  functional, 170
Griffiths–Kelly–Sherman inequalities, 127

Hamiltonian, 11
Hartree approximation, 126
heat bath, 97
Heisenberg fields, 162
Hermitian ordering, 49
Hilbert
  manifold, 66
  space, 62
Hilbert–Schmidt
  norm, 166
  operator, 165
history-integrals, 169
Hoelder's inequality, 2
homogeneous chaos, 157
homotopy group, 145

instantaneous sources, 76
interacting quantum fields, 157
interaction representation, 167
internal energy, 4
invariant integration, 164
Ising
  exchange energy, 126
  model, 3
isonormal measure, 165
isospectrality, 93
Ito integral, 40
Ito's formula, 24

Jacobi
  equation, 149
  field, 149
  matrix, 150
  theta function, 147
Jacobian, 176

Kac theorems, 53
Kelvin's formula, 176
Killing vectors, 183
Klein–Gordon equation, 36
Kolmogorov structure, 121
Kondo effect, 130
Konteweg-de Vries
  equation, 89
  flow, 89
Kuo's theorem, 61

Lagrangian, 34
  method, 170
Langevin equation, 97
Laplace
  asymptotic formula, 18
  method, 155
  operator, 137
Laplace–Beltrami operator, 34
lattice approximation, 137
Lax–Phillips structures, 120
Lebesgue
  dominated convergence theorem, 13
  measure, 6
Legendre elliptic function, 94
Levi–Civita connection, 63
Levy neighbourhood, 20
Lie group, 43
localized magnetic moments, 124
log concave functions, 1
loop
  expansion method, 184
  integral, 39
Lorentz
  frame, 158
  group, 70
  inner product, 72
  metric, 184

magnetic field, 37
many-body problems, 124
Markov
  field equation, 139
  processes, 18
  property, 70
mass shell, 187
Maxwell's equations, 77
McShane belated integral, 62
melting ice, 85
Minkowski space, 75
modified theta functions, 12
monomers, 53
multiminicomputer, 84

neutral scalar field, 136
Newtonian mechanics, 97
non-polynomial Lagrangian, 182
normal distribution, 69

operator-valued distribution, 160

parametrix, 47
parastatistics, 148
partition function, 47
path integral, 35
Pauli susceptibility, 130
permutation group, 148
perturbation theory, 159

phase transition, 126
Plancherel theorem, 73
plasma, 1
plastic, 59
Poincaré group, 71
Poisson summation formula, 12
polaron path-integral, 126
polymer
    melts, 59
    molecules, 53
Proca field, 69
propagator, 35

quantized
    distributions, 160
    scalar field, 160
quantum
    electrodynamics, 165
    field theory, 69
    gravity, 181
    mechanics, 34
    statistics, 147

radiation gauge, 78
Radon–Nikodym derivative, 60
random
    distribution, 136
    field, 136
    flight ensemble, 54
    walk, 94
reflection coefficient, 87
relativistic equations, 164
Riemann integral, 145
Riemannian
    manifolds, 43
    metric, 61
    spaces, 43
Ricci identity, 44
rubber elasticity, 55
Ruse's invariant, 47

saddle-point, 150
scattering
    theory, 87
    transformation, 166
Schrödinger equation, 34
Schwinger functions, 136
Schwinger's source theory, 69
SO (2), 43
SO (3), 43
Sobolev
    manifolds, 65
    norm, 71
    space, 79
specific heat, 4
spin, 147
spin glass, 57

spinor, 43
    curvature, 44
stationary phase, 144
statistical
    mechanics, 4
    operator, 124
statistics, 147
stochastic
    differential equation, 62
    geometry, 85
    integral, 36
Stokes's theorem, 39
Stone's thorem, 160
Stratonovich integral, 48
string model, 99
Sturm–Liouville
    equation, 150
    problem, 89
superconductivity, 124
susceptibility, 57

tempered distributions, 74
tensor extension, 42
tetrahedral
    lattice, 84
    topology, 84
thermodynamic potential, 125
time-ordering, 125
Toda lattice, 96
trace-class operators, 175
transition metal, 127
transmission coefficient, 87
Trotter product formula, 7

ultraviolet divergencies, 182
unitary
    dilations, 76
    dynamics, 159

Van der Waals gas, 127
Van Vleck determinant, 35
variational equation, 166
variation of constants, 167
vector-valued processes, 101
Veronoi tesselation, 85
viscosity, 54

wave equation, 34
weak
    distribution, 66
    solution, 161
Weyl
    mapping, 122
    operator, 162
    system, 160
Wick
    ordering, 142
    power, 162

Subject Index

Wiener
  density, 61
  integral, 8
  manifold, 60
  measure, 7
  paths, 7
Wightman
  axioms, 141
  distributions, 136

Wigner lattice, 1
Wilson theory, 126
WKB expression, 35

Yang–Mills gauge, 44